The Waste Market

The Waste Market

Institutional Developments in Europe

Edited by

E. Dijkgraaf
SEOR-ECRi, Erasmus University Rotterdam, Rotterdam, The Netherlands

and

R.H.J.M. Gradus
Faculty of Economics and Business Administration, VU University Amsterdam, Amsterdam, The Netherlands

Springer

E. Dijkgraaf
SEOR-ECRi
Erasmus University Rotterdam
P.O. Box 1738
3000 DR Rotterdam
The Netherlands

R.H.J.M. Gradus
Faculty of Economics and Business Administration
VU University Amsterdam
De Boelelaan 1105
1081 HV Amsterdam
The Netherlands

ISBN 978-1-4020-8710-3 ISBN 978-1-4020-8711-0 (eBook)

Library of Congress Control Number: 2008930933

Printed on acid-free paper

9 8 7 6 5 4 3 2 1

springer.com

Contents

Contributors

Germà Bel
Universitat de Barcelona and ppre-IREA, Dep. de Política Econòmica,
Av. Diagonal, 690 08034 Barcelona, Spain, e-mail: gbel@ub.edu

E. Dijkgraaf
SEOR-ECRi, Erasmus University Rotterdam, P.O. Box 1738,
3000 DR Rotterdam, The Netherlands, e-mail: dijkgraaf@few.eur.nl

X. Gellynck
Department of Agricultural Economics, Ghent University, Coupure links 653,
B-9000 Ghent, e-mail: xavier.gellynck@UGent.be

R.H.J.M. Gradus
Faculty of Economics and Business Administration, VU University Amsterdam, De Boelelaan 1105, 1081 HV Amsterdam, The Netherlands, e-mail: rgradus@feweb.vu.nl

B. Melenberg
Tilburg University, K 615, P.O. Box 90153, 5000 LE Tilburg, The Netherlands,
e-mail: b.melenberg@uvt.nl

H. Ohlsson
Department of Economics, Uppsala University, P.O. Box 513, SE-751 20 Uppsala, Sweden, e-mail: henry.ohlsson@nek.uu.se

R.J. Sørensen
Department of Public Governance, Norwegian School of Management, Nydalsveien 37, 0484 Oslo, Norway, e-mail: rune.sorensen@bi.no

P. Verhelst
Department of Agricultural Economics, Ghent University, Coupure links 653,
B-9000 Ghent, e-mail: Pieter.Verhelst@UGent.be

Chapter 1
Introduction

E. Dijkgraaf and R.H.J.M. Gradus

1.1 Introduction

In 2004 Elbert Dijkgraaf finished a PhD-thesis 'Regulating the Dutch waste market' at the Erasmus University Rotterdam. It was interesting that not much is published about the waste market, although it is a very important sector from an economic and environmental viewpoint. In 2006 we were participants at a very interesting conference on Local Government Reform: privatization and public-private collaboration in Barcelona organized by Germà Bel. It was interesting to notice that researchers from Spain, Scandinavian countries, the UK and the USA were studying this issue as well. From this we brought forward the idea to publish a book about the waste market. Because of its legal framework we want to focus on Europe.

In this chapter we give an introduction to this book. In the next paragraph we present a short overview of the waste collection market. Since 1960 the importance of the waste sector has increased substantially both in the waste streams and the costs of waste collection and treatment. Furthermore, we discuss policy measures to deal with these increases and give an overview of the different measures in EU-countries. In the last paragraph we present different chapters of our book.

1.2 Empirical Update of the Waste Collection Market

The Dutch case provides a nice example why studying the waste market is interesting from an economic point of view. The quantity of waste in kilograms per Dutch inhabitants has more than doubled in the last 90 years (see Fig. 1.1). After

✉ E. Dijkgraaf
SEOR-ECRi, Erasmus University Rotterdam, P.O. Box 1738, 3000 DR Rotterdam, The Netherlands
e-mail: dijkgraaf@few.eur.nl

R.H.J.M. Gradus
Faculty of Economics and Business Administration, VU University Amsterdam, De Boelelaan 1105, 1081 HV Amsterdam, The Netherlands
e-mail: rgradus@feweb.vu.nl

E. Dijkgraaf, R.H.J.M. Gradus (eds.), *The Waste Market*,

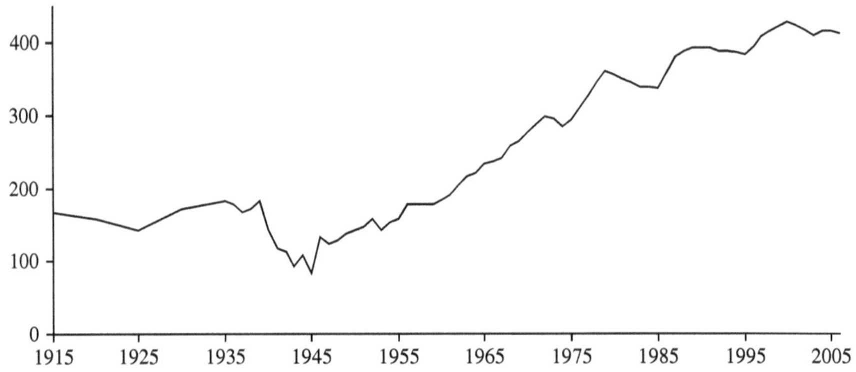

Fig. 1.1 Quantity of waste in kg per inhabitant, The Netherlands

2000 the level has stayed more or less stable on a level between 410 and 430 km per inhabitant.[1] As will be shown in this book (Chapter 8) the use of unit-based systems in some parts of the Netherlands in the last years is an explanation for this. Key question is than whether wider application of this system might result in much lower levels of waste and decreasing costs for citizens.

The increase in waste quantity and the changes in waste management policy resulted in a sharp acceleration of collection costs (see Fig. 1.2). In 1972 a Dutch

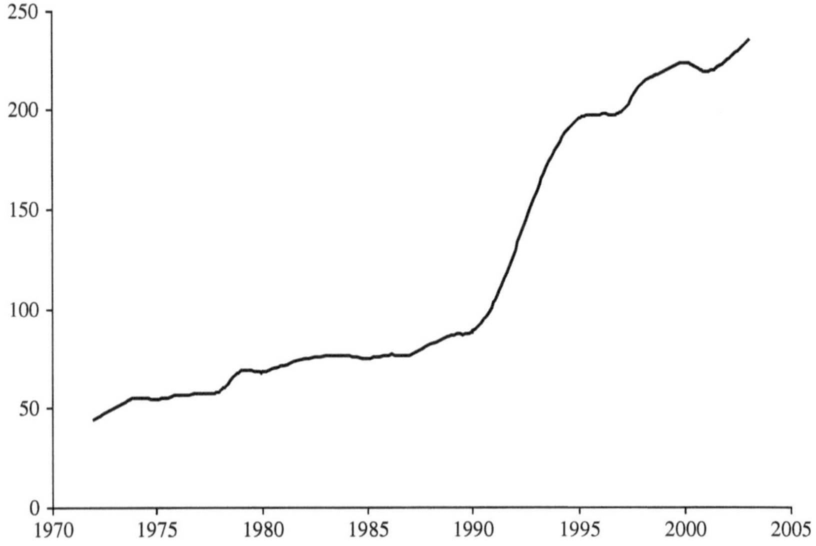

Fig. 1.2 Real costs waste in euro per household, The Netherlands

[1] Note that this includes the types of waste analyzed in this book. This is the sum of unsorted waste, recyclable waste (glass, paper, textiles) and compostable waste (vegetable, fruit and garden waste). We do not analyze demolition waste, chemical waste and other special types of waste.

household paid 44 euro per year on average for the collection and treatment of waste. In 1990 the real costs were already two times as high, while in 2003 a household paid more than five times as much.

This sharp rise in costs is not only a consequence of the increase of waste quantity, but also of the increased use of more expensive treatment options. Especially, the introduction of a landfill tax in 1996 and the introduction of more expensive incineration methods were important in this respect.[2] Due to this sharp increase several policy measures such as increasing private involvement of waste collection and unit-based pricing has been introduced with the goal to minimize waste collection costs. Main objective for this book is to learn from these experiences to get hold on possibilities to compensate the sharp rise in costs. This asks for an evaluation of the relation between costs and different policy measures such as privatization and variable charging as will be done in this book.

In 2002 a EU-study was published giving an overview of the extent of private involvement of refuse collection and variable charging (see Hogg, 2002). In Table 1.1 the differences between the EU-countries and Norway are summarized.

Table 1.1 Private sector involvement and variable charging in EU-countries

Country	Private sector involvement collection	Variable charging
Austria	50%	Widespread, usually on basis of volume
Belgium	Frequently	Widespread and increasing
Denmark	80%	10% of authorities, usually weight-based; some charge for additional bags
Finland	Municipalities dominant	Volume-based charging on residual waste
France	50%	14% of population, mostly volume based, some weight-based
Germany	Limited role	Widespread, by volume, amount of waste and sometimes frequency
Greece	Limited role	No variable charging
Ireland	40%	Being piloted, tagged bags, volume and weight based
Italy	46%	Will be compulsory, tags and average weight
Luxemburg	Some	Compulsory, mostly volume-based, some weight-based
Netherlands	38%	21% of municipalities, volume and volume/frequency are most common
Norway	10-15%	56% voluntary and 18% mandatory
Portugal	Limited role	No variable charging
Spain	56%	No variable charging
Sweden	60%	About 5% of municipalities, mainly based on size, some on weight
United Kingdom	50%	No variable charging permitted by law

Source: Hogg (2002, Table 1), Kipperberg (2007) and authors.

[2] This book only studies waste collection. See Dijkgraaf (2004) for an overview of the waste treatment market (landfilling and incineration) and options to reduce costs also in this market.

From Table 1.1 it is quite clear that private sector involvement differs across countries. In some countries, especially some Scandinavian countries such as Denmark and Sweden, the level of private sector involvement is large. In other countries such as Austria, the Netherlands and the UK, the level is less than fifty percent. In a third group, with countries such as Greece and Germany, there is a limited role for the private sector in waste collection. Although the results from this EU-study should be interpreted with caution as other studies give a slightly different picture and data can be outdated, it indicates that there is no firm one way trend in all EU-countries. Therefore, it is interesting to study the effects and reasons for privatization. Furthermore, variable charging or unit-based pricing also varies in its extent across EU-countries. Especially in the southern parts of Europe, except Italy, variable charging has not been implemented. In other countries such as the Benelux and the Scandinavian countries different forms of variable charging based on weight, volume and frequency are becoming more widespread and therefore are studied in several chapters in this book.

1.3 Description and Purpose of the Book

In this book we analyze the waste collection market in different EU-countries. In the previous paragraph we gave an empirical update of the waste market, which has changed considerable in the last thirty years. Especially in high densely populated countries or regions as the Netherlands, Belgium, Catalonia and the Stockholm area, waste management policies have resulted in an acceleration of waste collection and treatment costs.

In the following chapters several policy measures are discussed with the goal to generate more insight in the available policy options to reduce these costs. First, the cost advantage of contracting out refuse collection is analyzed. Second, evidence is presented for the incidence of contracting out related to this cost advantage. Key question is why private provision is not accepted as a best-practice and how this relates to issues like assuring enough control for municipalities, ideology, pressure groups and the dynamics of the market structure. This is done for the Netherlands (Chapters 1, 2, 3 and 7), Sweden (Chapter 4), Norway (Chapter 5) and Spain (Chapter 6). Third, the effects of unit-based pricing and other policy measures to decrease waste generation are studied based on experience in the Netherlands (Chapter 8) and Belgium (Chapter 9). Finally, Chapter 10 discusses items for future research.

Chapter 2 discusses the possible cost savings of contracting out refuse collection in the Netherlands. The findings indicate that similar to foreign econometric studies cost savings of approximately 15–20% apply to the Netherlands (for an overview see Domberger and Jensen, 1997). However, it should be noticed that contracting out is more important than the ownership issue because the difference in cost advantage between private and public firms is very small. In addition, compared with the existing literature it is shown that different production technologies apply to internal municipal waste collection units and external refuse collection firms. Different cost functions have to be estimated for the sub-samples. Using different production techniques it is shown that out-side firms such as private or public firms can make more use of scale economies than municipal service or cooperation.

There are some reasons to doubt the cost-advantage result of private contracting out in a dynamic perspective (see also Bel and Warner 2006). Contracting out refuse collection is a dynamic process typically converging from a competitive market structure to a monopolistic one. Even though the bidding process may have been competitive, the market becomes a bilateral monopoly just after awarding the contract. Contracted firms will try to keep control over the contract by means of anticompetitive behavior against rivals. This might explain why the use of private collectors seems relatively low, despite the estimated cost advantages at short term. In 2006 in the Netherlands, 38% percent of the municipalities used private firms. So, more than 60% of the municipalities has public provision (public firms, municipal cooperation or municipal collection service). For the United Kingdom and Sweden similar pictures can be drawn (see Chapter 4 for Sweden and Szymanski (1996) for the United Kingdom).

It is, therefore, important to study political economy factors that induce or deter privatization. In Chapter 3 an explanation is sought for the reservations of Dutch local authorities toward privatization. Based on theoretical insights the choice is modeled between private and public provision of refuse collection on the one hand and the choice between in-house and out-house provision on the other. Data are available for nearly all Dutch municipalities in 1998. Evidence is found that the higher the number of inhabitants the less likely it is that municipalities will privatize. In addition, large transfers by the central government and strong interest by public unions discourage privatization. Interestingly, the results with respect to the political variables are much weaker. For out-house provision the over-all results are in line with privatization. Compared to earlier studies also more general models are estimated. Although the same qualitative results are found for parametric and semi-parametric models, strong statistical evidence is found that a parametric specification is too inflexible. In addition, semi-parametric models are more capable to investigate spatial models.

In Chapter 4 it is shown that municipalities did not choose the least-cost alternative using Swedish 1989-data. In other words, cost differences did not affect producer choice in Sweden. Interestingly, other variables as ideology, the influence of pressure groups and legal constraints did not fit the data as well. In addition, the dummy-variable approach which captures the difference between public and private production is rejected and therefore a pooling model is used, which also corrects for selection bias. In that case it is shown that public production is 6% cheaper than private production. This is interesting because the cost advantage of public versus private production is the reverse, although the difference is rather small.

A more general finding in the political-economy literature is that there do not seem to be many ideological biases influencing politician's decision. The decision of the municipality is often pragmatic and not ideological. It seems that some politicians are reluctant to privatization simply because they do not foresee relevant cost savings in the longer term. A possible explanation is that initial savings given by privatization are diminished over time and the previous chapters seriously doubt the positive link between privatization and cost savings. Also in a recent meta-analysis of all published empirical refuse collection studies Bel and Warner (2006) show that there is little evidence for such a link.

In Chapter 5 the refuse collection process in Norway and the possible cost savings are discussed. Interestingly, in Norway only 10–15% of the municipalities use a private collector. It seems that privatization faces political opposition of the affected constituencies, which seems to imply that Norway is the less market oriented of the Nordic countries. However, to take advantage of economies of scale Norwegian municipalities are more willing to cooperate with other municipalities, which is the case for more or less half of the municipalities. In this chapter the internal governance structure is analyzed into more detail. It is shown that dispersed public ownership impairs inefficiency. Local governments that cooperate with neighbors to provide refuse services have costs that are 10% higher than those municipalities that supply the service single-handedly. In addition, if ownership is measured by the Herfindahl index of ownership concentration, estimates suggest that an increase in ownership concentration from 0 to 1 will reduce costs by 6% and an increase in the number of owners from 1 to 6 will increase costs by about 5%.

Chapters 6 and 7 describe the structure of the refuse collection market in more detail. Chapter 6 starts with analyzing the Spanish refuse collection market. Based on a 2003-sample for Spain, it is shown that the level of private provision (63%)[3] is higher than in some other European countries. Therefore, the Spanish system of municipal cooperation is described as it combines local grip with private provision. This system and the relatively low municipality size might explain the relatively high figure of private provision. With respect to the refuse collection market a dual market is faced. One single Spanish company concentrates more than 50% of the population served by private enterprises. Measured in the so-called Herfindahl index for Spain it is 0.33, which is a very high level. However, for the sub-sample of small municipalities this Herfindahl index and thereby the concentration is much lower. Nevertheless, Spanish private companies have significant market power in 2002 and it is, therefore, possible that the cost reductions of contracting out can not be sustained over time.

In Chapter 7 it is shown that the Dutch market for private refuse collection is highly concentrated as well. Moreover, it is shown that in highly concentrated provinces competition is weak, which results in barriers for local governments to effectively obtain benefits from contracting-out. However, according to our estimates this is only the case for private firms. It seems that the price behavior of public firms is not influenced by market concentration and in low concentrated provinces, where public firms are active, competition is strengthened. The importance of public firms is also put forward, if the increase of prices is related to the institutional form chosen. For the public firm dummy a significant negative effect is found meaning that prices go down if a public firm is chosen. However, the level of concentration does not

---

[3] This figure also includes so-called mixed-firms (see Table 6.1 of Chapter 6).

influence the cost saving effect of public firms, but the change in concentration does. Nevertheless, it seems an effective way to organize day-to-day operations under private commercial law rules, whereas the government retains control over strategic decisions. An important policy implication of this chapter is that local governments should be cautious with privatization of public firms.

In this book we analyze also unit-based pricing (UBP) as an instrument to lower refuse collection costs. With a UBP system the waste collection tariff depends on the amount of waste citizens produce. The idea is that this introduces an incentive for citizens to reduce the amount of waste compared with the generally used flat rate system. It is shown that unit-based pricing is much more effective in cost reductions of the costs of refuse collection than contracting out. From an environmental point of view this is important as well, because unit-based pricing systems are effective in reducing unsorted, the most environmentally unfriendly waste stream, and in stimulating recyclable waste.

In Chapter 8 the effects are estimated of four unit-based pricing systems (weight-, bag-, volume- and frequency-based) on waste collected using a panel data set for all Dutch municipalities. More than 20% of the Dutch municipalities had implemented such a system in 2000, while in 2005 this was already more than 30% (Dijkgraaf & Gradus, 2008). Unit-based pricing is shown to be very effective in reducing solid waste, composted waste and in increasing recyclable waste such as paper, glass and textile. If the estimations are corrected for differences in environmental activism between municipalities the effects are still large but significantly lower. The performance of bag- and weight-based systems is equal and compared with the frequency- and volume-based systems these two performs much better with a reduction of total waste of one third. This is interesting, as administrative costs are substantially lower for the bag-based system. Furthermore, unit-based pricing systems have no effect on the amounts of waste collected in surrounding municipalities.

Finally, the issue of illegal dumping, one of the adverse effects of the introduction of unit-based pricing systems is discussed as well. However, studying the effects of introducing a weight-based system in the Dutch municipality of Oostzaan Linderhof, Kooreman, Allers, and Wiersma, (2001) state that illegal dumping is virtually non-existent. According to them, the monitoring system in Oostzaan, with fines for illegal dumping, appears to be very effective in terms of deterrence. Moreover, another explanation for the absence of illegal dumping is that a small municipality such as Oostzaan has a large degree of social control. In general, the high population density of the Netherlands would suggest a low level of illegal dumping. This is confirmed by the lack of clear anecdotal evidence despite the large number of municipalities with unit-based pricing. Nevertheless, it shows that there is an important relation of unit-based pricing in relation to other policy variables as well.

Therefore, it is interesting to study the Flemish region of Belgium, where the authorities in the 'implementation plan household waste 2003–2007' assessed a broad policy mix (Chapter 9). It is shown that besides pecuniary incentives service level and measurements stimulating prevention and waste reduction are effective in reducing household solid waste. Instruments to reduce waste can be divided in three groups: pecuniary incentives; service level and measurements stimulating prevention and waste reduction. Also specific characteristics of the community determine the amount of waste generated. The chapter analyses whether findings in the literature on effectiveness of policy measures are valid for Belgium, specifically for the Flemish region. Multiple regression analysis identifies those measurements having the greatest impact on household solid waste. An income elasticity is found of 0.33. Also the provided service level has a significant impact. Pecuniary incentives are effective instruments in reducing waste, with a price elasticity of −0.14. Furthermore, a higher percentage of direct costs, directly attributable to waste services, borne by households, reduces waste. A consequent implementation of the 'polluter pays' principle proves to be effective.

Finally, Chapter 10 provides issues for future research.

References List

Bel, G., & Warner, M. (2006). *Local privatisation and costs: a review of the empirical evidence.* Working paper University of Barcelona.

Dijkgraaf, E. (2004). *Regulating the Dutch waste market.* PhD-thesis, Erasmus University Rotterdam.

Dijkgraaf, E., & Gradus, R. H. J. M. (2008). *Environmental activism and dynamics of unit-based pricing systems.* Working Paper, Vrije Universiteit Amsterdam.

Domberger, S., & Jensen, P. (1997). Contracting out by the public sector: theory, evidence and prospects. *Oxford Review of Economic Policy, 4,* 67–78.

Hogg, D. (2002). *Costs for municipal waste management in the EU.* Final report to Directorate General Environment, European Commission, Brussels.

Kipperberg, G. (2007). A comparison of household recycling behaviors in Norway and the United States. *Environmental & Resource Economics, 31,* 215–235.

Linderhof, V., Kooreman, P., Allers, M., & Wiersma, D. (2001). Weight-based pricing in the collection of household waste: The Oostzaan case. *Resource and Energy Economics, 23,* 359–371.

Szymanski, S. (1996). The impact of compulsory competitive tendering on refuse collection services. *Fiscal Studies, 17,* 1–19.

Warner, M., & Bel, G. (2008). Competition or monopoly? Comparing privatization of local public services in the US and Spain. *Public Administration,* 86, 1–13.

Chapter 2
Cost Savings of Contracting Out Refuse Collection in The Netherlands

E. Dijkgraaf and R.H.J.M. Gradus

Abstract This chapter discusses the possible cost savings of contracting out refuse collection in the Netherlands. Our findings indicate that similar to foreign econometric studies cost savings of approximately 15–20% apply to the Netherlands. Moreover, compared with the existing literature we show that different production technologies apply to internal municipal waste collection units and external refuse collection firms. Different cost functions have to be estimated for the sub-samples. Though significant cost savings exist on contracting out waste collection, households will not experience these cost savings on a one to one basis. Private refuse collection firms must pay VAT while public entities are exempted. Thus, the fiscal system hinders a more pronounced role for private refuse collection firms.

Keywords Collection · cost estimation · chow stability test · pooling · VAT

2.1 Introduction

Contracting out tasks like refuse collection, building cleaning, catering and vehicle maintenance has become an important measure to improve efficiency within the public sector. There is much evidence that contracting out certain public services can imply an efficient provision of services well adapted to needs and reduces the costs to tax payers. In an overview article Domberger and Jensen (1997) show that contracting out suggests cost savings in order of twenty percent without sacrificing the quality of service provided for a number of government services.

✉ E. Dijkgraaf
SEOR-ECRi, Erasmus University Rotterdam, P.O. Box 1738, 3000 DR Rotterdam, The Netherlands
e-mail: dijkgraaf@few.eur.nl

R.H.J.M. Gradus
Faculty of Economics and Business Administration, VU University Amsterdam, De Boelelaan 1105, 1081 HV Amsterdam, The Netherlands
e-mail: rgradus@feweb.vu.nl

E. Dijkgraaf, R.H.J.M. Gradus (eds.), *The Waste Market*,

In this chapter, we focus on the effects of contracting out refuse collection. A number of empirical studies are published on the effects of different institutional forms on performance in the waste collection market. The studies estimate the effects of private collection (or contracting out) by estimating a cost function. Generally, these studies show considerable cost savings, if refuse collection is contracted out.[1]

Kitchen (1976) estimates a cost decrease of $ 2.23 per capita when private firms collect household waste with data for 48 Canadian municipalities. Observations of 340 public and private firms in the USA, Stevens (1978) indicate a cost decrease of 7% to 30% due to contracting out. The magnitude of the effect depends on the size of the municipality. Pommorehne and Frey (1977) study refuse collection in Switzerland and again the private sector comes up with lower costs that amounted to 20%. Domberger et al. (1986) published a study on the effects of contracting out household refuse collection in the United Kingdom. Making use of a data set with 610 observations for 305 municipalities, they concluded that there are cost savings of 22% for contracting out to private companies and 17% for contracting in-house. Szymanski and Wilkins (1993) and Szymanski (1996) have confirmed the results, based on an extension (in years) of this database. Ohlsson (1998) reports comparable efficiency gains of contracting out for Sweden. Bosch, Predaja, and Suárez-Pandiello (2000) analyze Spanish data for 73 municipalities in Catalonia. They pointed out that the framework for which the service is provided is more relevant than the public private dichotomy. In a recent contribution Reeves and Barrow (2000) pointed out cost savings of around 45% in Ireland.

Though studies are performed for different countries, a study in the Netherlands is missing. We try to fill the gap and show that results of other studies are confirmed if we use comparable estimation techniques. Furthermore, we extend these studies in two directions. First, with the exception of Stevens (1978) all cited studies pool observations of waste collection units with respect to institutional forms to estimate the effects of contracting out. With this pooled data set a cost function is estimated and the coefficient of the included institutional dummy reveals the effect of different institutional forms. It is, however, questionable if this pooling is acceptable. Chow (1960) states that: 'Often there is no economic rationale in assuming that two relationships are completely the same' (p. 591). In other areas of economics Chow stability tests are used frequently, see e.g. Apergis, Papanastasiou, and Velentzas (1997), Lai (1994) and Loomis (1989). The most important application of the Chow stability test is to check for the Lucas critique in time-series. However, checking for different types of models with cross-sectional databases can be important as well.

[1] Some studies only compare the average cost for private versus public collection on the basis of ratio analysis, see e.g. Savas (1977, 1981) and McDavid (1985) or Data Envelopment Analysis, see e.g. Cubbin, Domberger, and Meadowcroft (1987). However, these methods fail to account for the effects in changes of other variables. By estimating a cost function, institutional effects but also other factors as the frequency of collection and density of the infrastructure can be taken into account. Therefore, we rely on this method in this chapter.

A priori it is not sure whether external refuse collection firms (outside firms) apply the same waste collection technology as internal municipal waste collection units (inside firms). Outside firms handle the collection process from a different perspective while organizational goals also differ. Moreover, differences in municipality size can lead to different collection techniques. For instance, bigger cities have more opportunities to make use of scale economies. If production techniques are not identical, pooling can lead to biased coefficients. Therefore, if pooling is not justified, different cost functions have to be estimated for each sub-sample. The omission of the checks on the validity of pooling in the mentioned studies may lead to biased estimated effects of contracting out on performance. From a policy perspective, it is important that estimations of possible cost savings are accurate.

Secondly, compared with previous studies more emphasis is put on the fiscal system. Due to the Dutch fiscal system there is a disincentive for contracting out. Even though we can estimate significant cost savings when waste collection is contracted out, households will not experience these cost savings on a one to one basis. In the Netherlands private collection firms have to pay VAT while public firms are exempt. Countries such as the United Kingdom and Denmark have a compensating system, in that local authorities are tax-neutral toward contracting outside or inside. Thus, the current fiscal system in the Netherlands renounces the role for private collection firms.

2.2 Effects of Tendering: Estimations for The Netherlands

Although many foreign econometric studies on effects of contracting out refuse collection have been published, such estimations are not available in the Netherlands. This section is an attempt to fill this gap by estimating a cost function, making use of a representative data set for Dutch municipalities. To make the results comparable the applied technique in this section corresponds with the studies cited in the previous section. The Chow stability test is applied in the next section.

2.2.1 Method

On the basis of previous research (see e.g. Stevens, 1978) the following standard equation is estimated:[2]

$$C = \alpha_1 Q + \alpha_2 I + \alpha_3 D + \alpha_4 F + \alpha_5 G + \alpha_6 P + \alpha_7 V + \alpha_8 O + \alpha_9 \qquad (2.1)$$

The driving forces behind the (logarithm of) total collection cost per municipality (C), include a number of variables.[3] First, the number of pick-up points (Q) is expected to determine part of the total cost. This reflects on the cost, which a collection

[2] Based on a Cobb-Douglas production technique and minimization of a total cost function.

[3] No price variables for the different inputs are included, because no reason exists ex ante why factor prices would differ between municipalities.

unit has to make by the number of stops. Secondly, the time spent at the pick-up stop (more bags or bins) can determine total cost. The number of inhabitants per pick-up point (I) approximates these costs. A third driving force is the time to arrive at the different pick-up points. The density variable, surface per pick-up point (D), approximates this. Fourth, the frequency of collection (F) is expected to have influence on total collection cost and is therefore included. Furthermore, the percentage of glass (G), paper (P) and vegetable, fruit and garden waste (V) separately collected is included in the estimations.

Furthermore, we include a dummy for the institutional form in which waste is collected (O). Main difference of the institutional form is whether waste is collected by the municipality itself or outside. Within this category we can discriminate between two types on the basis of ownership, i.e. public and private. Public outside collectors are a combination of municipalities for which waste is collected by an other municipality and municipalities that formed an independent public organization. Given the division of institutional forms, the basic model is tested whether the ownership of the outside collection service does matter.

Expected signs are positive for the number of pick-up points, inhabitants per pick-up point, surface per pick-up point and collection frequency and negative for the institutional dummy's, while signs of the coefficients for the percentage collected glass, paper and vegetable, fruit and garden waste are undetermined a priori.

2.2.2 Data

To collect data 120 municipalities were approached in the period November 1996-April 1997. These municipalities were selected at random from 646 Dutch municipalities. A total of 85 municipalities have responded to this inquiry, a response rate of 71%.[4] The 85 municipalities responded to an inquiry on the collection of waste in 1996. The resulting database was checked on consistency of answers and the reliability was checked by spot checks on key answers.

Of the 85 municipalities 41 collect their waste not inside, but trough an outside organization (see Table 2.1). Of the 41 outside firms, 13 were public independent organizations while 3 municipalities collect the waste through an other municipality. The remaining 25 municipalities collected the waste through a private collection firm.

Total cost per municipality is measured by multiplying the refuse collection rate(s) by the total number of households. Total cost is diminished by handling cost by multiplying cost per ton with tons recycled (glass and paper), composted vegetable, fruit and garden waste) and disposed (incineration and dumping).

---

[4] In 1996 four municipalities were absorbed by another, 31 municipalities did not participate in this inquiry.

Table 2.1 Data description

	Average	Maximum	Minimum	St.dev.
Total cost (million euro)	1.6	20.5	0.1	2.5
Pick-up points (number)	16,386	267,000	400	3,0618
Inhabitants (per pick-up point)	4.0	64.7	1.6	8.1
Density (km^2 per pick-up point)	11	93	1	15
Frequency (>1 per week, dummy)	0.19	1.00	0.00	0.39
Glass (%)	3.2	11.1	0.0	3.0
Paper (%)	6.6	29.7	0.0	7.5
VFG (%)[a]	28.4	47.4	0.0	9.9
Outside collection (dummy)	0.48	1.00	0.00	0.50
Private outside collection (dummy)	0.29	1.00	0.00	0.46
Public outside collection (dummy)	0.19	1.00	0.00	0.39

Note: [a] VFG = vegetable, fruit and garden waste

2.2.3 *Fiscal Aspects*

A lot of attention has been drawn to the distortionary aspects of taxation for all kind of commodities (see Atkinson and Stiglitz, 1980). For the central question in this chapter taxation can also be crucial. The fiscal regime distorts the decision process in the Netherlands with respect to public versus private waste collection (see Wassenaar, 2001). Private refuse collection is faced with a VAT rate of 19%, while public organizations are exempted from VAT. Therefore, a municipality in the Netherlands is biased toward inside production, because then refuse collection is exempted from VAT.

A possibility to resolve this inequality could be to assess public refuse collection as a business activity and thus tax them with VAT. This policy has been introduced to public companies such as telecommunications. However, taxing refuse collection by municipalities is not allowed according to EU laws. The other extreme, introducing a VAT exemption for enterprises is also not allowed.

The ministry of Finance has been working on a system to create a VAT compensation fund for public waste collectors (Wassenaar and Gradus, 2001). In line with a system already working in United Kingdom, all VAT a municipality has to pay will be refunded. In that case a municipality that decides to contract out the waste collection to a VAT liable firm will be compensated for the VAT the firm has to pay. Thus, contracting out decisions by a municipality are no longer distorted by the VAT difference between public and private firms.

The difference in fiscal treatment cannot be neglected for the Dutch data set for a proper analysis. The municipality cost for private companies are 19% higher compared to public companies. However, the costs for a private company are 19% lower and in this respect the cost data are corrected.[5] Thus, the VAT component is subtracted from the total cost for private firms.

[5] A the cost data are for the fiscal year 1996, the VAT correction is based on the tariff of that year (17.5%).

2.2.4 Results

Results for the basis model are presented in the first column of Table 2.2. The F-statistic shows that the equation is significant, while the high (adjusted) R^2 indicate that the explained variation is high. All coefficients have the expected sign. T-statistics are not corrected for heteroscedasticity as the White test (White, 1980) could not reject the homoscedasticity hypothesis for all estimations with 95% confidence.

The number of pick-up points has a significant impact on the total collection cost. A Wald test of coefficient restrictions (Pindyck and Rubinfeld, 1991) does not falsify the constant returns to scale hypothesis. This result confirms earlier results from Reeves and Barrow (2000), Collins and Downes (1977) and Hirsch (1965), while Stevens (1978) found also constant returns to scale for the large cities. Decreasing returns to scale were found by Bosch, Predaja, and Suárez-Pandiello (2000) and Domberger et al. (1986) and increasing returns to scale in Szymanski and Wilkins (1993), but coefficients were very close to one. Kitchen's (1976) inverted U-shaped average cost curve result was not confirmed since inclusion of a quadratic term was falsified with an F-test on 95% confidence.

Table 2.2 Estimation results cost functions

		Dummy for outside collection	Dummy for outside and private outside collection
Pick-up points	ln	1.052 (20.90)	1.052 (20.81)
Inhabitants per point	ln	1.004 (12.34)	1.007 (12.29)
Density (km^2 per point)	ln	0.009 (0.23)	0.010 (0.24)
Frequency	dummy	0.174 (2.07)	0.177 (2.10)
Glass	%	0.019 (1.41)	0.018 (1.36)
Paper	%	−0.008 (−1.40)	−0.007 (−1.25)
VFG	%	−0.010 (−2.26)	−0.010 (−2.06)
Private and public outside	dummy	−0.163 (−2.18)	−0.134 (−1.44)
Private outside	dummy	–	−0.051 (−0.50)
Constant		4.13 (6.96)	4.10 (6.84)
R^2		0.93	0.93
F-value		132.30	116.48
Log likelihood		−11.36	−11.22
White (prob. Homoscedasticity)		0.41	0.40
Number of observations		85	85

Note: Below coefficients are t-statistics. VFG = vegetable, fruit and garden waste

The number of inhabitants per pick-up point, the pick-up frequency and the percentage of collected vegetable, fruit and garden waste have a significant impact on total cost. If the number of inhabitants per pick-up points increases with 1%, the total cost will rise with the same percentage. A higher pick-up frequency leads to 19% higher cost. Total cost decrease if more vegetable, fruit and garden waste are collected. It may be due to a scale effect as vegetable, fruit and garden waste is collected on a one bin per household while the number of bins per household is fixed.

The dummy for outside collection is significant. On average outside provision leads to 15% lower total cost.[6] In the second column the hypothesis is tested whether private outside collection does have an effect on total cost above that of outside provision. The negative coefficient implies that on average private collection is 5% cheaper than public collection. However, the basic model, without an ownership dummy, is not rejected on the basis of a Log-likelihood-ratio test (test statistic is 0.28). Furthermore, the dummy for ownership is not significant, while the coefficient for outside provision in the extended model does not differ from the basic model (using a Wald-test). Thus, we can conclude that the choice between outside and inside provision is more important than the ownership of the collection service. Competition seems to have more effects than the ownership issue. This is consistent with the literature (see Domberger and Jensen, 1997).

Compared to Domberger et al. (1986) and Szymanski (1996) effects of changing institutional forms are somewhat lower but of the same order. Maybe competition in the Netherlands is somewhat less stringent since the private firms are not numerous. Three firms with only some small local private collection firms dominate private collection in the Netherlands.

An important result from our findings is that the difference in fiscal treatment between private and public 'firms' hampers tendering on the waste collection market.[7] Tendering to a private firm will not result in significant effects on tariffs paid by households. Dutch local governments are free to decide either to collect the waste by themselves or to tender the job. However, from January 2003 a VAT compensation fund is present for public waste collectors. According to our results this initiative will lead to a decrease in social cost of waste collection.

---

[6] Calculated as e^x-1, where x is the value of the estimation for the dummy for outside provision.

[7] The corrections made because of the difference in tax treatment (17.5%) could be too high as public collectors can not deduct paid VAT on inputs. This paid VAT is part of the price consumers pay for the collection of waste. However, inputs with a VAT obligation are very low in total cost. For example total cost for collection trucks are only about 10% of total collection cost. This would result in a 1.75% point lower difference in effective VAT rates between public and private waste collectors. Moreover, the obligation for private firms to pay profit tax would diminish this difference as capital cost rise. Regressions with a 1% point lower effective VAT rate for private firms show only very small differences in coefficients for the institutional dummy's. Even a 10% point lower effective VAT rate for private firms results in a significant cost decrease if waste is collected by an outside firm.

2.3 Robustness of Results

As Ganley and Grahl (1988) make clear the results for institutional dummies can be influence by specific observations that perform much better or worse than expected. Therefore, we tested whether our result for the outside dummy remains robust when we skip municipalities with much lower or higher cost than expected. By iteration we excluded municipalities with a higher deviation of predicted to real cost than 30%. The outside dummy remains significant (but now even at 99%), while the coefficient remains robust.

An other point to investigate is whether the estimations depend on extreme small or big municipalities. Therefore, we tested whether a dummy for very big or small municipalities should be added to our basic model. Using a Log-likelihood-ratio test the basic model is not rejected.

Szymanski and Wilkins (1993) test for sample selection bias. They have two reasons to suspect that sample selection bias could be a problem for their estimations. First, they estimate a cost function for a data set including different years while the response rate in 1988 was significant lower than in other years. This may be due to the introduction of compulsory competitive tendering in that year. Moreover, they suspect that authorities which performed a successful competitive tender were certainly keen to report, whereas an inefficient controlled authority did not likely to report (p. 117). As we do not have an indication that comparable problems exist in the Netherlands, we assume that sample selection bias is not a crucial problem. Furthermore, Szymanski and Wilkins (1993) found that there model without corrections for sample selection bias is not rejected.

Stevens (1978) tested for the validity of pooling the different municipalities in one sample. She concludes that different estimations have to be made for a few municipality size classes, but that pooling of the private and public collection firms was valid. Also Ganley and Grahl (1988), in a reaction to Domberger et al. (1986), emphasize to make a difference between urban and rural municipalities. Domberger, Meadowcroft, and Thompson, (1988) state in their reply that the included dummy for rural versus urban municipalities solves this problem. However, they did not check explicitly the validity of pooling the observations.

Chow (1960) made clear that testing for the validity of pooling observations is possible (see also Fischer, 1970). As unjust pooling of observations can lead to biased estimated coefficients this validity check is also necessary. Therefore, we checked the validity of pooling the observations for the Dutch data set with respect to municipality size and the different institutional forms, making use of the Chow test.[8]

Testing for the hypothesis that breakpoints exist with respect to small, mid-size and large municipalities reveal that this hypothesis cannot be rejected

[8] Toyoda (1974) and Schmidt and Sickles (1977) showed that the Chow test for equality of regression coefficients is not robust to heteroscedasticity. Then other tests can be applied (see e.g. Thursby, 1992) Fortunately, the homoscedasticity hypothesis is not rejected for our estimations.

Table 2.3 Chow breakpoint test cost function Netherlands

Breakpoint between rest versus:	No breakpoint hypothesis		
	F-statistic	Probability	Conclusion
Public and private outside collection	2.98	0.01	breakpoint
Private outside collection	1.93	0.07	no breakpoint
Public versus private outside collection	1.98	0.03	breakpoints
< 20, 000 inhabitants	3.58	0.00	breakpoint
< 40, 000 inhabitants	0.30	0.96	no breakpoint
> 20, 000 and < 40,000 inhabitants	2.02	0.03	breakpoints

(see Table 2.3). The impossibility to reject the breakpoint hypothesis with respect to municipality size could be due to the relative inflexible Cobb-Douglas form of the production function. However, testing for size breakpoints with a more flexible translog form holds the same conclusions.[9] Moreover, a breakpoint hypothesis with respect to the different institutional forms cannot be rejected. The probability that no breakpoints exist for all three organization forms is less than 5%.[10] This means that different cost functions must be estimated for the three institutional forms. For reasons of both types of breakpoints, our estimates in the previous section could be biased.

Combination of the two different breakpoint tests results in 6 sub-sample estimations. As our sample includes only 85 municipalities the estimations would become meaningless. Therefore, we follow a three-step approach. First, we take into account the effects of pooling the three sub-samples related to institutional form by estimating three equations. Secondly, we test these equations for the validity of pooling the observations with respect to municipality size. Third, we make some calculations based on nonparametric methods to estimate the effect of institutional form on cost.

Table 2.4 reveals the effects of sub-sampling on the basis of the different institutional forms. Comparing the coefficients for the estimated equations clearly reveals that they are significantly different. Apparently, inside, public and private outside waste collectors have a different production technology. These results give an indication that outside firms can make more use of economies of scale. This is not surprising as municipal waste collectors are bounded on their borders. Outside waste collection firms are more flexible as they can combine the collection of different municipalities. The number of inhabitants per pick-up point is significant in the 'inside' equation, while they have no significant effect on the cost of the different outside firms. This applies also for the relative part of vegetable, fruit and garden refuse in total waste.

[9] The translog cost function has exactly the number of parameters required for a flexible functional form, see e.g. Diewert (1987).

[10] Although a breakpoint is rejected at the 95% level for private collection versus other institutional forms, a breakpoint between private outside collection, public outside collection and inside collection could not be rejected.

Table 2.4 Estimation results cost function, different institutional forms

		Inside	Private outside	Public outside
Pick-up points	ln	1.103	1.044	0.964
		(15.86)	(8.28)	(12.21)
Inhabitants per point	ln	1.100	-1.333	-2.047
		(12.49)	(-0.47)	(-1.94)
Density (km^2 per point)	ln	-0.000	0.109	-0.015
		(-0.00)	(0.87)	(-0.16)
Frequency	dummy	0.137	0.209	0.109
		(1.50)	(1.03)	(0.34)
Glass	%	0.014	-0.017	0.015
		(0.67)	(-0.64)	(0.54)
Paper	%	-0.004	-0.010	0.002
		(-0.49)	(-0.96)	(0.28)
VFG	%	-0.012	-0.010	0.004
		(-2.13)	(-0.91)	(0.37)
Constant		3.593	5.265	7.259
		(4.59)	(3.65)	(4.54)
R^2		0.91	0.80	0.98
F-value		61.78	14.52	109.55
White (probability homoscedasticity)		0.22	0.55	0.66
Number of observations		44	25	16

Note: Below coefficients are t-statistics. VFG = vegetable, fruit and garden waste

We tested the three estimated equations for the validity of pooling the observations with respect to municipality size, again with a Chow test. Table 2.5 summarizes the results. Each equation was tested for breakpoints, the number of tests only limited by the number of observations. Reported is the maximal F-statistic found per equation. For the equations for private outside and inside waste collectors the Chow breakpoint test reveals that the no-breakpoint hypothesis could not be rejected. Therefore, we conclude that pooling with respect to municipality size was valid for these cases. Due to the low number of observations, the equation for public outside collectors could not be tested for breakpoints.

While the samples are now homogenous for institutional form, it is not possible to include a dummy for this variable in the estimations. Nonparametric comparison however can give an indication of possible cost differences between the samples. The estimated equations can be used to predict the development of cost when the institutional form is changed. Total cost for municipal collectors if they are contracted

Table 2.5 Chow breakpoint test cost function, institutional sample

Estimation:	Inhabitants:	Maximal F-statistic	Probability (no breakpoint)
Private outside	19,000	2.17	0.13
Public outside[a]	na	na	na
Inside	27,500	1.70	0.14

[a] Breakpoint test is not available due to low number of observations

Table 2.6 Estimated cost increases and institutional change (% total cost)

From outside	to	inside	17.2
From private outside	to	inside	19.3
From public outside	to	inside	14.0
From inside	to	private outside	−14.8
From public outside	to	private outside	3.4
From inside + public outside	to	private outside	−9.9
From inside	to	public outside	−13.9

out can be predicted with the estimated equation for private collectors, making use of the known variables for municipal collectors.

Predictions using the estimated equations based on sub-samples confirm the cost decrease effect of changing the institutional form to a more market related direction. Contracting out the inside collection to a private firm would yield an average cost decrease of 14.8% (see Table 2.6). This is almost exactly what we found with the pooled estimation for the basic model. If the institutional form of inside waste collectors is changed to public outside the estimated cost decrease is 13.9%, only 1% lower than we found earlier. Of interest is the prediction for bringing outside firms inside. Apparently municipalities that collect waste by means of contracting outside have a very good reason for doing that as the predicted average cost increase is large.

2.4 Conclusions

While empirical research on the effects of changes in institutional form on the waste collection market for the Netherlands is missing, this chapter fills in the gap. Our results confirm the results of earlier studies, i.e. contracting out refuse collection results in lower cost of 15–20%. Moreover, we can conclude that the choice between outside and inside provision is more important than the ownership of the collection service. Competition seems to have more effects than the ownership issue.

The statistical analysis indicates that waste collectors in smaller, medium and big municipalities have different production technologies. This also applies for different institutional forms. As more flexibility exist with respect to combining the collection of different municipalities, outside firms can make more use of economies of scale.

The fiscal system in the Netherlands hinders a more profound role for private waste collection as households will not benefit of the possible cost decreases. The burden of higher taxes for private firms counteracts the efficiency improvements. A VAT compensation fund would further stimulate the role of private waste collection. The current actions taken by the Ministry of Finance to correct the VAT difference between public and private firms are necessary to stimulate a fair choice between the real advantages and disadvantages of contracting out.

Acknowledgments We thank R.C.G. Haffner, M. Varkevisser for comments on an earlier version and K.G. Berden for his work on the database. This chapter was published as Dijkgraaf E, Gradus RHJM (2003) Cost savings of contracting out refuse collection. Empirica 30:149–161.

References List

Apergis, N., Papanastasiou, J., & Velentzas K (1997). The credibility of policy announcements: Greek evidence. *Applied Economics, 29,* 699–705.

Atkinson, A. B., & Stiglitz, J. E. (1980). *Lectures on public economics*. London: McGraw-Hill.

Bosch, N., Predaja, F., & Suárez-Pandiello, J. (2000). Measuring the efficiency of Spanish municipal refuse collection services. *Local Government Studies*, 26, 71–90

Chow, G. C. (1960). Tests of equality between sets of coefficients in two linear regressions. *Econometrica*, 28, 591–605

Collins, J. N., & Downes, B. T. (1977). The effects of size on the provision of public services: The case of solid waste collection in smaller cities. *Urban Affairs Quarterly,* 12, 333–347.

Cubbin, J., Domberger, S., Meadowcroft, S. (1987). Competitive tendering and refuse collection: Identifying the sources of efficiency gains. *Fiscal Studies*, 8, 49–58.

Diewert, W. E. (1987). Cost functions. In: J. Eatwell, M. Milgate, & P. Newman (Eds.), *The new Palgrave, a dictionary of economics* (pp. 690–696). London: MacMillan.

Domberger, S., & Jensen, P. (1997). Contracting out by the public sector: theory, evidence and prospects. *Oxford Review of Economic Policy,* 4, 67–78.

Domberger, S., Meadowcroft, S., & Thompson, D. (1986). Competitive tendering and efficiency: The case of refuse collection. *Fiscal Studies*, 7, 69–87.

Domberger, S., Meadowcroft, S., & Thompson, D. (1988) Competition and efficiency in refuse collection: A reply. *Fiscal Studies*, 9, 86–90.

Fisher, F. M. (1970). Tests of equality between sets of coefficients in two linear regressions: An expository note. *Econometrica*, 38, 361–366.

Ganley, J., & Grahl, J. (1988). Competition and efficiency in refuse collection: A critical comment. *Fiscal Studies*, 9, 80–85.

Hirsch, W. Z. (1965, February) Cost functions of an urban government service: refuse collection. *Review of Economics and Statistics*, 47, 87–92.

Kitchen, H. M. (1976). A statistical estimation of an operating cost function for municipal refuse collection. *Public Finance Quarterly*, 4, 43–72.

Lai, C. C. (1994) Market structure and inter-industry profit differences in Taiwan. *Pakistan Development Review*, 33, 147–163.

Loomis, J. B. (1989). Test-retest reliability of the contingent valuation method: A comparison of general population and visitor responses. *American Journal of Agricultural Economics*, 71, 76–84.

McDavid, J. (1985, Sept/Oct). The Canadian experience with privatising residential solid waste collection services. *Public Administration Review*, 45, 602–608.

Ohlsson, H. (1998). Ownership and Production Costs: Choosing between Public Production and Contracting Out. Uppsala University working paper 6.

Pindyck, R. S., & Rubinfeld, D. L. (1991). *Econometric models & economic forecasts*. New-York: McGraw-Hill.

Pommorehne, W. W., & Frey, B. S. (1977). Public versus private production efficiency in Switzerland: A theoretical and empirical comparison. *Urban Affairs Annual Review*, 12, 221–241.

Reeves, E., & Barrow, M. (2000). The impact of contracting out on the costs of refuse collection services: The case of Ireland. *Economic and Social Review*, 31, 129–150.

Savas, E. (1977). Policy analysis for local government: Public vs. private refuse collection. *Policy Analysis*, 3, 49–74.

Savas, E. (1981). Intracity competition between public and private delivery. *Public Administration Review*, 41, 46–52.

Schmidt, P., & Sickles, R. (1977). Some further evidence on the use of the Chow test under heteroscedasticity. *Econometrica*, 45, 1293–1298.

Stevens, B. J. (1978). Scale, market structure and the cost of refuse collection. *Review of Economics and Statistics*, 60, 438–448.

Szymanski, S., & Wilkins, S. (1993). Cheap rubbish? Competitive tendering and contracting out in refuse collection – 1981–1988. *Fiscal Studies*, 14, 109–130.
Szymanski, S. (1996). The impact of compulsory competitive tendering on refuse collection services. *Fiscal Studies*, 17, 1–19.
Thursby, J. G. (1992). A comparison of several exact and approximate tests for structural shift under heteroscedasticity. *Journal of Econometrics*, 53, 363–386.
Toyoda, T. (1974). Use of the Chow test under heteroscedasticity. *Econometrica*, 42, 601–608.
Wassenaar, M. C., & Gradus, R. H. J. M. (2001). *Contracting out: the importance of a level playing field* (Research Memorandum 0108). Erasmus University Rotterdam.
White, H. (1980). A heteroskedasticity-consistent covariance estimator and a direct test for heteroskedasticity. *Econometrica*, 48, 817–838

Chapter 3
Contracting Out Refuse Collection in The Netherlands

E. Dijkgraaf, R.H.J.M. Gradus and B. Melenberg

Abstract In this chapter we seek an explanation for the reservations of local authorities towardcontracting out. Although empirical evidence suggests that contracting out results in a significant cost decrease, a majority of Dutch municipalities provides for waste collection services themselves. Based on theoretical insights we model the choice between private, public, in-house, and out-house refuse collection. The models are estimated using a database comprising nearly all Dutch municipalities. We find evidence that the number of inhabitants, the transfer by central government, and interest group arguments are important explanations. Interestingly, ideology seems to play a minor role. Compared to earlier studies we estimate more general models. Although the same qualitative results are found for parametric and semiparametric models, we find strong statistical evidence that a parametric specification is far too inflexible. Differences between the parametric and the semiparametric marginal effects are substantial. Thus, more attention is needed for the implications of model specification.

Keywords Refuse collection · institutional choice · ideology · interest groups · semiparametric estimation

✉ E. Dijkgraaf
SEOR-ECRi, Erasmus University Rotterdam, P.O. Box 1738, 3000 DR Rotterdam, The Netherlands
e-mail:dijkgraaf@few.eur.nl

R.H.J.M. Gradus
Faculty of Economics and Business Administration, VU University Amsterdam, De Boelelaan 1105, 1081 HV Amsterdam, The Netherlands
e-mail: rgradus@feweb.vu.nl

B. Melenberg
Tilburg University, K 615, P.O. Box 90153, 5000 LE Tilburg, The Netherlands
e-mail: b.melenberg@uvt.nl

E. Dijkgraaf, R.H.J.M. Gradus (eds.), *The Waste Market*,

3.1 Introduction

There seems to be evidence that contracting out government services saves taxpayers money, and sometimes a lot of money, compared to public provision. In an overview, Domberger and Jensen (1997) show that contracting out a broad field of government services might result in cost savings in the order of 20% without sacrificing the quality of services provided.

Also Tang (1997), in a critical assessment of several studies, comes to the conclusion that the private sector is found to be more efficient in refuse collection, fire protection, cleaning services, and capital intensive waste-water treatment, while in sectors as water supply and railways the results are more mixed.

Especially, the cost savings of private refuse collection have been discussed at length in the literature. Kitchen (1976) estimates a cost decrease of Canadian $ 2.23 per capita when private firms collect household waste. Stevens (1978) arrives at a cost decrease between 7% and 30% due to contracting out for the USA, where the magnitude of the effect depends on the size of the municipality. Based on UK-data Domberger, Meadowcroft, and Thompson (1986) published a study on the effects of contracting out household refuse collection in the United Kingdom. They concluded that there are cost savings of 22% for contracting out to private companies. Szymanski and Wilkins (1993) and Szymanski (1996) have confirmed these results, based on an extension (in years) of this database. Dijkgraaf and Gradus (1997) show similar cost savings between 15% and 20% for the Netherlands, in case Dutch municipalities are contracting out refuse collection. Moreover, Ohlsson (1998) reports almost the same estimations for Sweden. Recently, Bosch, Pedaja, and Suarez-Pandiello (2000) presented Spanish data for 73 municipalities in Catalonia. They pointed out that the framework for which the service is provided is more relevant than the public private dichotomy. In a recent contribution Reeves and Barrow (2000) pointed out cost savings of around 45 % in Ireland.

Although the practice of contracting out refuse collection has become more popular, it is still less common than in-house provision. In the United Kingdom only 30% of the contracts for refuse collection is placed out-house (see Szymanski, 1996). According to Reeves and Barrow (2000), in Ireland in 39 % of the studied cases private providers were contracted to provide refuse collection. In the Netherlands 40% of the municipalities use private collectors for refuse. However, due to the fact that private collectors are especially active in small villages, only 20% of total tonnage is in private hands (see Dijkgraaf & Gradus, 1997). Only Ohlsson (1998) finds for the Swedish case that private provision is slightly more common than public provision.

Furthermore, a study by López-de-Silanes, Shleifer, and Vishny (1997) shows the reservations of local authorities toward contracting out. Based on data in 1987 and 1992 for 3042 counties for twelve services like water supply, landfills, libraries etc. only 25% of the services in 1987 and 35% in 1992 had been placed out-house. Moreover, in this chapter a nice empirical investigation of the mode of providing government services is given, where three leading theories (namely efficiency, political patronage, and ideology) are investigated. The evidence presented in this chapter indicates that clean government laws and state laws restricting county spending encourage privatization, whereas strong public unions discourage it. This suggests an important role played by political patronage and taxpayer resistance to government spending in the privatization decision.

In this chapter, we examine for the Netherlands the determinants of the provision mode of refuse collection. Data are available for 540 (i.e., almost all) Dutch municipalities. We find evidence for political patronage and the wealth of the local government as a ground for contracting out, but also the possible efficiency gain of contracting out plays a role. Moreover, we extend the existing literature by investigating more general specifications. Especially, the usually applied logit model seems too restrictive. Formal tests strongly reject the appropriateness of the logistic probability transformation. As alternative we use a semiparametric single index modeling approach, based on Ichimura (1993), where the probability transformation is left unrestricted. We find that the semiparametric single indices are comparable to the parametric analogues, but the probability transformations are quite different, implying that the logit specification might yield misleading predictions, particularly, when considering marginal effects.

The remainder of this chapter is organized as follows. In Section 3.2 we discuss the relevant theoretical issues. In Section 3.3 we describe the data we use. Section 3.4 contains the estimation results based on logit. In Section 3.5 we investigate the robustness of these results, by testing the logit specification, and by using a semiparametric alternative, based on Ichimura (1993). Section 3.6 concludes.

3.2 Theoretical Issues

Before we specify the data and the empirical results, it is worthwhile to discuss some theoretical issues concerning the contracting out decision (see also López-de-Silanes et al., 1997; Tang, 1997). As mentioned in the introduction, Dijkgraaf and Gradus (1997) show that Dutch municipalities might achieve cost savings between 15% and 20% in case of contracting out refuse collection. With lower service costs, one would expect that municipalities favor private collection. Indeed, 40% of the Dutch municipalities chose for the option to collect waste by a private firm. The question arises: why did the other 60% not choose this option as well?

Hart, Shleifer and Vishny (1997) argue that private contractors might fail to pursue goals that politicians want to attain. Especially, in circumstances such as health care and prisons, where politicians cannot write a complete contract that specifies exactly what contractors are supposed to do in all circumstances, it may not be straightforward to contract out. The logic suggests some potential efficiency benefits of in-house government services to ensure quality. However, it is not clear how important such benefits are for refuse collection. Hart et al. (1997, p. 1154) argue that in the case of refuse collection the damage to quality can be offset by a good contract, so that "private provision is superior". Nevertheless, according to a Dutch inquiry, such elements are still available and some municipalities put forward that quality is the reason for in-house provision (see NG magazine , 1998). A prediction following from this kind of reasoning is that the wealth of local government decreases the likelihood of contracting out. A poorer government is less likely to care about quality and is more interested in cost savings.

Related to these wealth arguments are the so-called output arguments. Some empirical insights suggests a linear relation between the cost of service and output (number of inhabitants, pick up points etc., see, for example, Domberger et al., 1986). However, especially for small municipalities this may not be true. Kitchen (1976) finds that the maximum scale in refuse collection occurs in cities of about 324,000 inhabitants. Stevens (1978) divides the sample into several subsamples. For small municipalities there is less evidence for this linear relation. Therefore, she finds increasing returns to scale, if the city population is less than fifty thousand and constant returns to scale if the city population is larger than fifty thousand. A prediction following from this kind of reasoning is that the number of inhabitants decreases the likelihood of contracting out. However, this relation may not be linear. Above a certain level there is less evidence that private waste collectors have more opportunities to combine the collection of different municipalities and thus to use scale effects as a cost decreasing mechanism.

An alternative view of the contracting out decision focuses on public choice theory (see Buchanan, 1987). This approach explains social behavior as the product of free choices of individuals. Self-interested politicians, bureaucrats and unions have a stake in in-house provision as they can use it as a status-enhancing feature. López-de-Silanes et al. (1997) argue that in the United States the main political factor favoring in-house provision seems to be the public employee unions. Moreover, the role of unions becomes more important and, therefore, in-house provision becomes more beneficiary if unemployment in a municipality is high.

The third theory stresses the importance of voter ideology. To evaluate this view, one should take into account voting patterns in different municipalities. Hereby, it is assumed that the contracting out decision is simultaneously determined by the degree of voters' anti-government sentiment. This laissez-faire sentiment is most visible in right-wing parties.

Finally, it is possible that the privatization decision in a particular municipality is related to what happens in other municipalities. For instance, Bivand and Szymanski (2000) find evidence for the UK that in the period before Compulsory Competitive Tendering (CCT) costs were spatially correlated across authorities, while following CCT this spatial correlation disappeared. To account for this effect, Bivand and Szymanski suggest that before CCT most local authorities evaluated the service costs by comparison with their local neighbors. Municipalities with a higher than average cost compared with the neighbors would choose the option of privatization. In addition, the decision of contiguous municipalities might affect the decision of a municipality via scale economy, especially when the municipality under consideration is small. Alternatively, one could argue that municipalities might take into account the decisions in some kind of reference group of municipalities, where the reference group consists of municipalities which are, for instance, comparable in size or in number of inhabitants.

However, contrary to the first three points, this fourth issue, interdependence between municipalities is much harder to quantify. Without knowledge of which municipalities influence which municipalities, the researcher will have to model such interdependencies him- or herself by modeling reference groups. However, as argued by Manski (1993) in the context of a linear demand equation for consumers with interdependencies between consumers, it is impossible to infer unknown reference groups on the basis of observed behavior: an informed specification of reference groups is a necessary prelude to an analysis of interdependent behavior. As such information is not available for our case estimation of the effects of interdependencies is not possible.[1]

3.3 Data

To test the theories about contracting out, a database is constructed with data on the different institutional forms of waste collection and variables representing the theories. The data on the different institutional forms is based on a 1998 census of the Dutch Association for Refuse and Cleansing Management (NVRD). Moreover, municipalities' characteristics are available from Statistics Netherlands (CBS). For 540 of the Dutch municipalities (96% of all municipalities) figures are available, see Table 3.1.

Table 3.1 Descriptive statistics database

Variables	Average	Maximum	Minimum	St. dev.
Private provision (%)	42	100	0	49
In-house provision (%)	28	100	0	45
Inhabitants (x 1000)	26	722	1	45
Inhabitants per hectare	6	63	0	8
Transfer from central government per inhab. (euro)	442	1727	118	113
Income per inhabitant (1000 Euro)	9	13	6	1
Unemployed per 100 inhabitant	3	6	1	1
Local civil servants per 100 inhabitant	11	16	8	3
Conservative Liberals (%)	16	52	0	9
Social Democrats (%)	16	49	0	9
Progressive Liberals (%)	8	34	0	7
Orthodox Protestants (%)	6	67	0	10
Green Left (%)	4	34	0	6
Extreme Right (%)	0	11	0	2
Local parties (%)	25	100	0	20

[1] Moreover, since our model is of a binary choice type, we would also have to deal with the problem of "coherency", when modeling interdependencies, see, for example Schmidt 1981 or Gourieroux 1980: the interdependency should be of a recursive type (one municipality may influence the other, but then not the other way around), since otherwise the model is not coherent, i.e., probabilities do not sum to one.

3.3.1 Institutional Forms

In general, three modes of provision are used in this dataset. The first mode is provision by a private firm (42%). The second and third mode are both by public ownership but differ with respect to autonomy of the collection service. The second mode occurs when municipalities collect the waste of their own citizens (28%). The waste collection service is in this case under direct control of the municipality council. The third mode occurs when another municipality or an external public organization (30%) collects the waste, so that the municipality council has less direct control on the waste collection service.

3.3.2 Output Variables

To check for the output arguments the number of inhabitants and population density (number of inhabitants per hectare) are included in the empirical setting. On average a Dutch municipality has 26 thousand inhabitants, while the largest city (Amsterdam) has 722 thousand inhabitants and the smallest municipality only 1 thousand. To check for scale economy the number of inhabitants squared is included as well. Moreover, the population density shows a high variation between municipalities, indicating that the transport distance between individual pick-up points varies.

3.3.3 Wealth Variables

The theory about the influence of wealth on contracting out suggests that budget constraints influence the trade-off between efficiency and social arguments. Hard budget constraints increase the likelihood of privatization. In the Netherlands the income of local government depends almost totally on the transfers by the central government. The freedom of Dutch municipalities to collect their own taxes is quite restricted. Therefore, we include as an explaining variable the transfer from central to local government per inhabitant. As the trade-off between efficiency and social arguments depends on the social characteristics of the inhabitants we include the average personal income in a municipality as a wealth variable as well. The hypothesis is that a municipality will weigh cost savings more if the inhabitants are poor.

3.3.4 Interest Group Variables

In the López-de-Silanes et al. study interest group variables are included for the number of public employee's or union membership and for the opportunity to purchase supplies from political allies (the so-called clean government variables). However, for the Netherlands clean government laws are dictated at a national level and, therefore, these data cannot be included. No data are available for the number of public employee's in a municipality. However, these data are available at a regional level and are, therefore, included.[2] Similar to López-de-Silanes et al., it is possible to include labor-market conditions as an approximation of interest group variables. In general, we should expect that in-house provision becomes more beneficiary if unemployment in a municipality is high. Therefore, the unemployment level is included in our estimations.

3.3.5 Political Variables

We include the fractions of the following parties, based on the local elections of May 1994[3]: green left, social democrats, conservative liberals, progressive liberals, orthodox Protestants, extreme right and local parties.[4] In the estimations the Christian democrats, who are in the middle of the political spectrum, are excluded.[5]

3.4 Estimation Results: Logit

We start our estimations with a standard logit analysis for two models.[6] In the first model, the choice between public and private provisions is estimated as dependent on a number of explaining variables. In the second model the choice between in-house and out-house provision is the dependent variable. In both models, all explaining variables are initially the same.[7] Thus, the basic model is:

$$P(Dep = 1|x) = \Lambda(\beta^T x) \tag{3.1}$$

where:

Dep: Dependent variable,
model 1: dummy with value 1 for municipalities with no private collection;
model 2: dummy with value 1 for municipalities with collection in-house;
and where x contains the following explanatory variables (next to a constant term):

[2] There are twelve provinces or regions in the Netherlands.
[3] There were new elections in May 1998.
[4] Green left: Groen Links + SP, social democrats: PvdA, conservative liberals: VVD, progressive liberals: D66, Christian democrats: CDA, orthodox Protestant: SGP + RPF + GPV, ultra right: CD and local parties: other parties. Combination of the parties is tested using a Log Likelihood test.
[5] In addition, we looked at municipality-level voting in the 1994-election for Parliament as alternative indicator of the electorate's ideological orientation. However, local elections seem to be the best means of predicting the probability of private contracting.
[6] The probit and the OLS results are extremely similar.
[7] An interesting extension would be to include the previous state of the dependent variable as an explanatory variable. However, such data are not available.

Inhabitants	Number of inhabitants (*10000);
Funds	Transfers from central government (Euro per inhabitant);
Income	Personal income (Euro per inhabitant);
Civil servants	Number of civil servants (per 100 inhabitants);
Unemployment	Number of persons with an unemployment benefit (per 100 inhabitants);
Conservative Liberals	percentage of total votes in a municipality;
Orthodox Protestants	percentage of total votes in a municipality;
Social Democrats	percentage of total votes in a municipality;
Progressive Liberals	percentage of total votes in a municipality;
Green Left	percentage of total votes in a municipality;
Extreme Right	percentage of total votes in a municipality;
Local Parties	percentage of total votes in a municipality.

To account for sufficient flexibility in terms of the number of inhabitants, we also decided to include the number of inhabitants squared (/1000). The parameter vector β contains the unknown parameters, and Λ represents the logit-transformation.

Results are given in Table 3.2. This table contains the estimated parameters, together with the estimated standard errors.

First, we discuss the **no-private** provision case.

Output variables It shows that scale effects are present. The estimated second order polynomial in terms of inhabitants is increasing up to its maximum at around 312,500 inhabitants, so that with an increasing number of inhabitants (up to this maximum) the probability of public provision (i.e., no private provision) increases.[8] The occurrence of scale effects makes public provision more likely. Furthermore, if the number of inhabitants per hectare increases the probability of public provision increases. Again scale effects are present.

Wealth variables As we expected, more transfers by the central government favors public provision, because less emphasis has to be given to cost savings investigations. Contrary to our prior, a higher income level in a municipality lowers the probability of public provision. However, the estimated coefficient is not significant.

Interest group variables Interesting are the results with respect to the interest group variables. The data give evidence for the prior that the number of public employees raises the probability of public provision. Also the number of unemployed persons raises the probability of public provision, although the coefficient estimate is not significant.

Political variables The results with respect to the political variables are much weaker.[9] Only local parties are against public provision in a significant way

---

[8] In the Netherlands only three cities have more inhabitants than this maximum, namely The Hague, Rotterdam, and Amsterdam.

[9] The insignificance of political variables may be sensitive to specification of these variables. Therefore, we experiment with a "left/right" variable. This "left/right" variable is constructed as follows: 8*Green Left + 7*Social Democrats + 6*Progressive Liberals + 5*Christen Democrats + 4*Local Parties + 3*Orthodox Protestant + 2*Conservative Liberals + 1*Extreme Right.

Table 3.2 Estimation results logit and Ichimura model waste collection

Variables	No-private (logit)	No-private (Ichimura)	In-house (logit)	In-house (Ichimura)
Constant	−1.72	–	−4.51	–
	(1.88)	–	(2.18)	–
Inhabitants	0.26	0.26	0.18	0.18
	(0.11)	–	(0.08)	–
Inhabitants squared	−4.16	−3.94	−2.45	−2.57
	(1.73)	(0.30)	(1.13)	(0.38)
Population density	0.08	0.07	0.03	0.05
	(0.03)	(0.03)	(0.02)	(0.01)
Fund	1.38	1.11	1.36	1.34
	(0.66)	(0.55)	(0.73)	(0.33)
Income	−1.61	−1.24	−0.67	−1.17
	(0.93)	(0.73)	(1.05)	(0.43)
Unemployment	0.02	0.02	0.10	0.04
	(0.02)	(0.02)	(0.03)	(0.01)
Civil servants	0.27	0.19	0.10	0.04
	(0.05)	(0.05)	(0.05)	(0.02)
Conservative Liberal (%)	−0.004	−0.01	0.0008	0.009
	(0.02)	(0.02)	(0.02)	(0.01)
Social Democrat (%)	−0.0003	−0.02	−0.04	−0.03
	(0.02)	(0.01)	(0.02)	(0.01)
Progressive Liberal (%)	0.008	−0.02	0.03	0.01
	(0.02)	(0.01)	(0.02)	(0.01)
Orthodox Protestant (%)	0.03	0.01	0.02	0.009
	(0.01)	(0.01)	(0.02)	(0.01)
Green Left (%)	−0.02	−0.04	−0.01	−0.01
	(0.02)	(0.02)	(0.02)	(0.01)
Extreme Right (%)	−0.05	−0.08	−0.18	−0.29
	(0.13)	(0.06)	(0.08)	(0.04)
Local party (%)	−0.02	−0.03	−0.03	−0.008
	(0.01)	(0.01)	(0.01)	(0.01)
Log likelihood	−281.52		−256.40	

Note: Estimated standard errors in brackets.

(compared to the Christian Democrats, who are the reference group). Probably, this can be explained by the anti-government sentiment by some of these local parties. From the other parties only the Orthodox Protestants are in favor of public provision in a significant way. This can probably be explained by the reserved attitude toward the role of market forces in these parties.

For **in-house provision** the over-all results are in line with no-private provision. The top of the polynomial in terms of inhabitants is now at around 367,000

A loglikelihood test was used to investigate whether the model with individual parties is preferred. We obtain as test result for the no-privatization case a value of 25.86 and for the in-house modeling a value of 27.52; since the test is asymptotically chi-squared-distributed with 6 degrees of freedom, the tests lead to rejection at the 1%-level, suggesting that the specification including different parties is preferred. Furthermore, the left/right variable is not significant at 10%.

inhabitants: thus, if the number of habitants increases (up to this maximum) then the probability of in-house provision increases. The effect of inhabitants per hectare, however, becomes insignificant. This also applies to the effect of transfers by the central government. For the number of unemployed persons the effect of in-house provision seems somewhat stronger and quite significant. The effect of income per inhabitant remains contrary to our prior, but the estimated coefficient is again not significant. The effect of the number of public employees is again positive and significant. This seems to be in line with the theory that interest group considerations are an important obstacle to out-house provision. In addition, the results for political variables are also here suggestive. The attitude of the social democrats and extreme right toward in-house provision turns out to be significant, whereas the effect of the other parties is insignificant (compared to the Christian Democrats).

3.5 Robustness of Results

The basic logit model presented in the previous section requires strong distributional assumptions to be valid. In particular, the assumption that the probability transformation is given by the logistic probability distribution Λ may be questioned. To investigate the validity of this assumption we tested it against a more general specification as proposed by Ruud 1984), and as used by Newey (1985) for constructing conditional moment tests.[10] Thus we test $H_0 : \gamma_1 = \gamma_2 = 0$ in

$$P(Dep = 1|x) = \Lambda(\beta^T x + \gamma_1(\beta^T x)^2 + \gamma_2(\beta^T x)^3) \tag{3.2}$$

using the test statistic proposed by Newey (1985), adapted to the logit specification. We obtain as test result for the no-privatization case a value of 6.16; since this test is asymptotically chi-squared-distributed with 2 degrees of freedom, the test leads to rejection of the logit specification at the 5%-level. In case of the in-house modeling the test result becomes much higher: 33.95; this means rejection of the logit specification at all usual significance levels.

Consequently, it makes sense to investigate alternative specifications, which require less severe distributional assumptions. One possibility is a fully nonparametric approach, but due to the curse of dimensionality this will not work in our case with only 540 observations. So, we restrict attention to semiparametric models. There are several possibilities available in the literature for application to the binary choice case. One possibility is the Maximum Score estimator proposed by Manski (1985), and turned into smoothed Maximum Score by Horowitz (1992). Although (Smoothed) Maximum Score requires very weak distributional assumptions it has some drawbacks: it has a lower rate of convergence than ordinary parametric estimators and it only allows one to estimate the index, but not the

[10] Newey (1985) considers the probit specification; however, the adaptation to the logit model is straightforward.

probability transformation. Another possibility are the single index models in which the probability that the binary dependent variable equals one given the covariates is equal to a single index of the covariates evaluated in an unrestricted (nonparametric) probability transformation:

$$P(Dep = 1|x) = H(\beta^T x), \tag{3.3}$$

where H is an unknown function that has to be estimated as well. There are several estimators available to estimate such single index models. For instance, Klein & Spady, (1993) provide a semiparametric efficient one. However, this estimator is quite hard to calculate in practice. We decided to use Ichimura (1993).[11] The estimator for β consists of solving the minimization problem

$$\hat{\beta} = Arg\min_b \sum_i (Dep_i - \hat{H}(b^T x_i))^2, \tag{3.4}$$

where $\hat{H}$ represents a nonparametric estimator for

$$P(Dep = 1|x) = E(Dep|\, x) = H(\beta^T x). \tag{3.5}$$

We estimate this latter conditional expectation using a kernel estimator with a standard normal Gaussian kernel. Since there is no optimality theory for the corresponding bandwidth, we have set it equal to the familiar rule of thumb $\hat{\sigma}\, n^{-1/5}$ with $\hat{\sigma}$ an estimate for the standard deviation of. $\hat{\beta}^T x$.[12] The resulting estimator for β has a normal limiting distribution whose asymptotic covariance matrix can straightforwardly be estimated. See Ichimura (1993) for further details.

Table 3.2 contains the estimation results for β, and Figs. 3.1 and 3.2 present the estimates for H, for the no-privatization and in-house case, respectively. Notice that in the single index model the constant term is not identified (therefore, set equal to 0). Also the scale is not identified; we have fixed the scale by normalizing the coefficient of the variable Inhabitants, equal to the corresponding estimated coefficient in the Logit model.

The estimation results in terms of β according to Ichimura are, at least qualitatively, quite comparable with those according to the logit specification. To investigate whether the results are also quantitatively the same, we considered the hypothesis that the coefficients of logit are (simultaneously) equal to the corresponding single-index coefficients of the Ichimura-specification. We tested this hypothesis by a Hausman-type test by using the difference of the vector of logit estimates and the corresponding Ichimura-estimates. The limit distribution of this difference can easily be obtained under the null hypothesis. The value of the resulting chi-square

[11] For other possibilities, see, for instance, Horowitz (1998).

[12] As a starting value for the iteration procedure we used the OLS-estimate for β, from which we also constructed the estimate for σ.

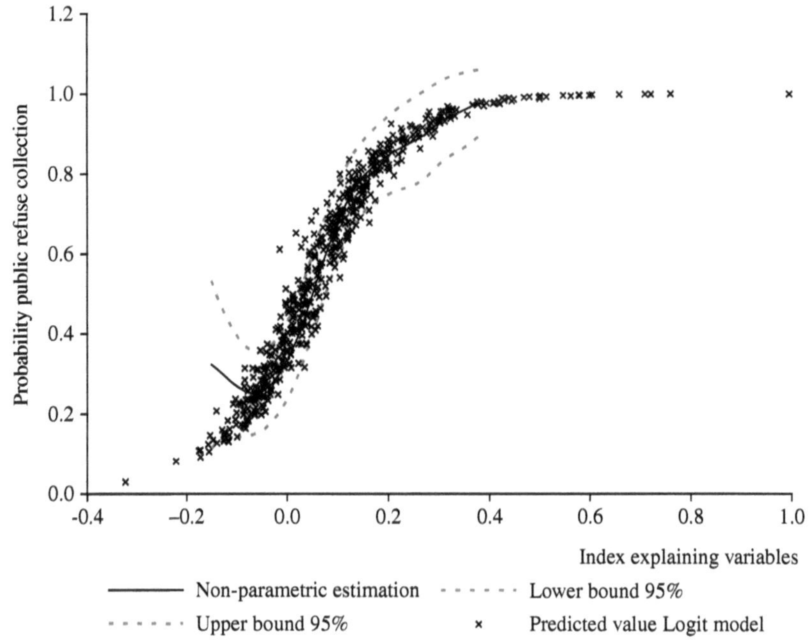

Fig. 3.1 Nonparametric (*solid-lines*) and logit (*dots*) estimation of choice between public and private refuse collection

test statistic turned out to be 1.75 in case of no-private collection, and 3.69 in case of in-house provision. Since, under the null hypothesis, the test statistic is asymptotically chi-square-distributed with 13 degrees of freedom (the number of coefficients, except the constant term and the normalized coefficient of inhabitants), we conclude that the results in terms of the single-index coefficients are also quantitatively the same.

Next, we turn to the estimated probability transformations. In Fig. 3.1 we plot the nonparametric estimate of the probability transformation in case of no-private-collection, together with 95% confidence intervals. In addition, we plot in the figure the corresponding predictions according to the logit model. From this figure we can conclude that the logit- and the Ichimura-specifications for most observations are not too far apart from each other. However, a non-negligible part of the predictions according to logit fall outside the 95% confidence interval, which can be seen as evidence that the logit model is misspecified, in line with the earlier rejection of the logit probability transformation. Moreover, for the lowest values of the single-index the results of Ichimura differ substantially from logit, although not significantly so. It seems that the probability transformation is not increasing over the whole range, a feature that cannot be captured by the logit-specification. In Fig. 3.2 we present the corresponding plot in case of in-house-provision. Again, we see that for many observations the logit- and Ichimura-specifications are reasonably close, but not as close as in case of no-private-collection: Over the whole range we see

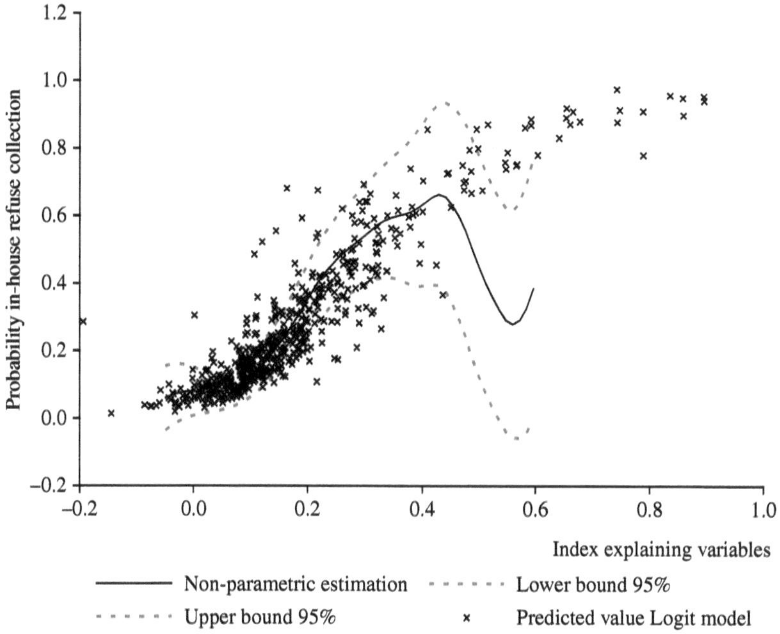

Fig. 3.2 Nonparametric (*solid lines*) and logit (*dots*) estimation of choice between in-house and out-house refuse collection

predictions according to logit falling outside the 95%-confidence band.[13] Moreover, for larger values of the single index, the Ichimura probability transformation is not increasing, but inversely hump shaped, a pattern that clearly cannot be captured by the logit probability transformation. Concluding, based on the overall evidence, the difference between logit and Ichimura is significant, in line with the earlier reported rejection of the logit probability transformation.

To investigate the consequences of the misfit by logit for a substantial part of our sample, we compare the prediction performances of the models, as well as the estimated marginal effects of changes in the covariates on the probabilities. First, Table 3.3 contains the prediction performances. For the sake of comparison, we also include in this table the naïve predictions without using any covariates. We predict the endogenous variable to be equal to one, if the predicted probability is at least a half; otherwise, we predict the endogenous variable as zero.

From this table we conclude that the prediction capabilities of both logit and Ichimura are quite comparable, and that Ichimura only slightly outperforms logit in both the no-private-collection and the in-house-collection cases. Of course, this is only a very rough comparison. To further illustrate how the predictions of both

[13] The number of inhabitants, population density and the share of local parties deviate for the municipalities outside the 95%-confidence band. Probably the Ichimura specification is especially superior for observations with special characteristics as this specification allows more flexibility.

Table 3.3 Percentage correct predictions of the various models

	No-private collection	In-house collection
Naïve prediction	0.58	0.72
Logit	0.7574	0.7685
Ichimura	0.7593	0.7796

specifications are in line with each other we plot in Figs. 3.3 and 3.4 the predictions according to logit against those according to the Ichimura specification. Fig. 3.3 contains the comparison for the no-private-collection case. The correlation coefficient between the predictions equals 0.97, but, particularly, at lower values of the single indices, we see a clear difference between both specifications, as already suggested by Fig. 3.1, but not reflected in Table 3.3. In Fig. 3.4 we consider the in-house-collection case. Here the correlation is much weaker than in Fig. 3.3. Indeed, the correlation coefficient is only 0.87, indicating that a blind use of logit may be misleading.

Thus, although the single index of logit corresponds quite closely to the single index according to Ichimura, the logit probability transformation is likely to be misspecified, due to its inflexibility, preventing it from fitting non-monotonic patterns. This might have implications for the quantification of the marginal effects of the covariates on the probabilities of no-private collection and in-house provision. To investigate this, we compare the estimated marginal effects of changes in the covariates on the predicted probabilities. We calculate these effects for each

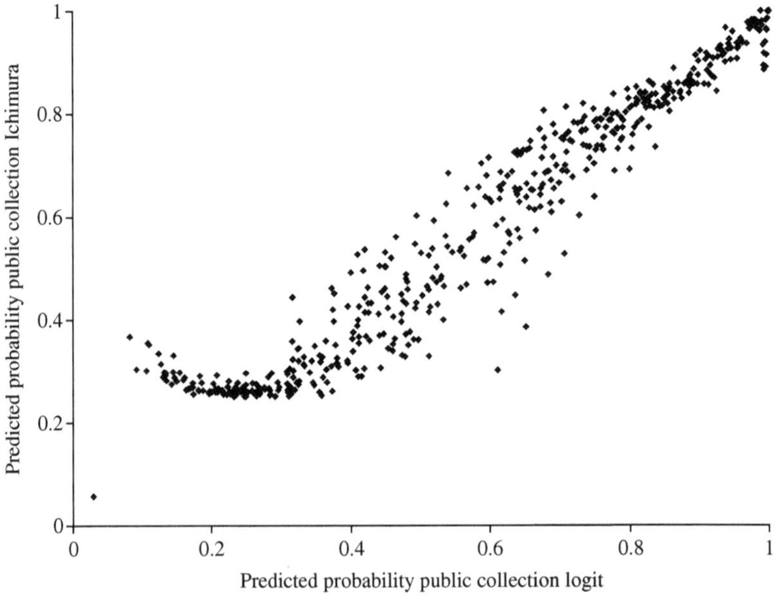

Fig. 3.3 Comparison predictions logit versus Ichimura (public versus private refuse collection)

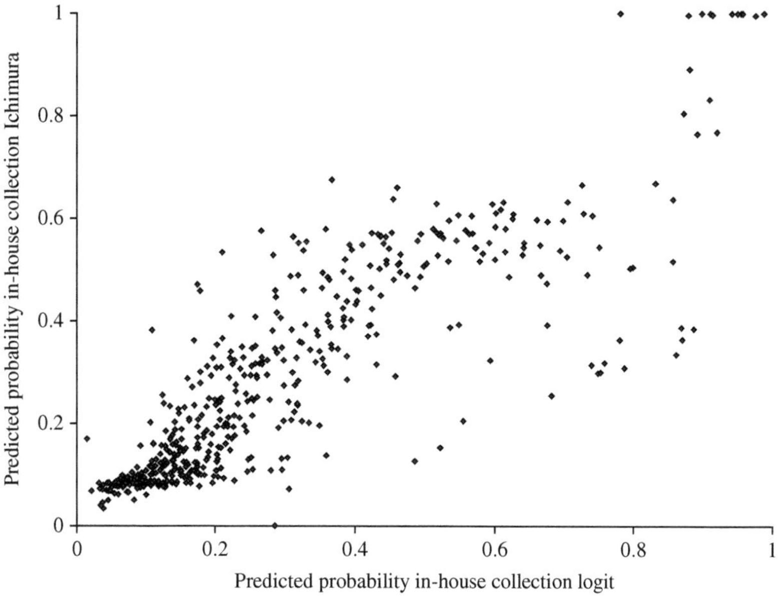

Fig. 3.4 Comparison predictions logit versus Ichimura (in-house versus out-house)

municipality in our sample, and then we average them over the sample. In this way we are measuring the (average) macro-effect of a marginal change in the covariates. Notice that our sample contains almost all Dutch municipalities, so that we are more or less dealing with the whole population. We include the standard deviations of the means to give an indication of the variability of the calculated effects, and we calculate the average of the absolute differences per municipality between the two models, to see how close the effects are. Table 3.4 contains the results for no-private-collection and Table 3.5 presents the results for in-house-collection.

Looking first at Table 3.4 (no-private-collection), we see in case of, for instance, the output variables (inhabitants or inhabitants per hectare) that the calculated average macro effects are quite comparable between the two specifications. However, in both cases, the average absolute differences are quite large compared to the average marginal effects, indicating that on the individual municipality level the models yield substantial differences, which, on an aggregate level, are averaged out. We also see that the variability in the logit marginal effects is much smaller than the variability in the Ichimura marginal effects, which, of course, is a consequence of the imposed monotonic logit probability transformation, as opposed to the non-monotonic Ichimura probability transformation. In case of the wealth variables and the interest group variables we have a similar story. Looking at the political variables, we see that in some cases the differences between both models are, at least qualitatively, substantial, although the magnitudes of the marginal effects are quite small.

Table 3.4 Marginal effects no-private-collection

Variable	Logit	Ichimura	Abs. Difference
Inhabitants	0.0437	0.0484	0.0269
	(0.0182)	(0.0425)	
Population density	0.0133	0.0135	0.0072
	(0.0053)	(0.0117)	
Funds	0.2413	0.2165	0.1101
	(0.0967)	(0.1871)	
Income	-0.2815	-0.2420	0.1223
	(0.1127)	(0.2091)	
Unemployment	0.0037	0.0035	0.0018
	(0.0015)	(0.0030)	
Civil servants	0.0471	0.0377	0.0191
	(0.0189)	(0.0325)	
Conservative Liberals	-0.0007	-0.0023	0.0020
	(0.0003)	(0.0020)	
Social Democrats	-0.0001	-0.0035	0.0038
	(0.0000)	(0.0030)	
Progressive Liberals	0.0014	-0.0030	0.0045
	(0.0006)	(0.0026)	
Orthodox Protestants	0.0047	0.0022	0.0026
	(0.0019)	(0.0029)	
Green Left	-0.0033	-0.0082	0.0065
	(0.0013)	(0.0071)	
Extreme Right	-0.0081	-0.0172	0.0131
	(0.0032)	(0.0148)	
Local Parties	-0.0041	-0.0062	0.0040
	(0.0016)	(0.0053)	

Note: Standard deviations in brackets

Turning next to Table 3.5 (in-house-provision), we see that the differences are now more substantial. For instance, in case of the output variables the estimated marginal effect according to Ichimura is between 1.6 (inhabitants) and 2.6 (inhabitants per hectare) times as large as the corresponding effect according to logit. In case of the wealth variable income per inhabitant the estimated negative marginal effect of income per inhabitant in case of Ichimura is even almost three times as large as in case of logit. The corresponding average absolute differences are also quite substantial. Similarly to the no-private-collection case, the logit marginal effects again show much less variability than the Ichimura marginal effects.

Concluding, we can state that, although the logit single index seems to be appropriate, the logit probability transformation seems to be too inflexible, producing, at least in the in-house provision case, average marginal effects whose magnitudes may be quite incorrect, and resulting in both the no-private collection and the in-house provision cases in an accuracy which may be quite misleading. By applying a semi-parametric specification, this inflexibility of the logit probability transformation can easily be circumvented.

Table 3.5 Marginal effects in-house-collection

Variable	Logit	Ichimura	Abs. Difference
Inhabitants	0.0257	0.0426	0.0314
	(0.0110)	(0.0432)	
Population density	0.0051	0.0136	0.0121
	(0.0022)	(0.0149)	
Funds	0.2092	0.3378	0.2596
	(0.0890)	(0.3702)	
Income	-0.1029	-0.2934	0.2659
	(0.0438)	(0.3215)	
Unemployment	0.0258	0.0109	0.0083
	(0.0067)	(0.0120)	
Civil servants	0.0157	0.0122	0.0084
	(0.0067)	(0.0123)	
Conservative Liberals	0.0001	0.0022	0.0025
	(0.0001)	(0.0025)	
Social Democrats	-0.0063	-0.0066	0.0045
	(0.0027)	(0.0072)	
Progressive Liberals	0.0042	0.0029	0.0022
	(0.0018)	(0.0032)	
Orthodox Protestants	0.0037	0.0024	0.0020
	(0.0016)	(0.0026)	
Green Left	-0.0018	-0.0034	0.0028
	(0.0008)	(0.0038)	
Extreme Right	-0.0270	-0.0721	0.0642
	(0.0115)	(0.0790)	
Local Parties	-0.0041	-0.0020	0.0024
	(0.0018)	(0.0022)	

Note: Standard deviations in brackets

So far, we considered no-private-provision and in-house-collection separately. However, one might argue that there may be some ordering present: at level 0 one can consider full privatization; at level 1 there is public provision, but not in-house; and at level 2 there is full in-house collection. Such an ordering may be modeled by a single index model as well. However, this only makes sense if the two indices, when estimating the choices no-private-collection and in-house-provision separately, are (more or less) the same. Therefore, we also considered the hypothesis that the vectors of coefficients of these two indices are equal. We tested this hypothesis by means of a Hausman-type test based on the difference between the two Ichimura-estimators, after appropriate scaling.[14] The resulting chi-square test statistic yielded as value 36.5, which results in strong rejection of the hypothesis of equal indices, since the critical value of a chi-square distribution with 13 degrees of freedom equals 22.36 (at 5%). We concluded that the modeling of the mentioned ordering by means of a single index is likely to yield a misspecified model, even if modeled semiparametrically. Therefore, we did not investigate this possibility further.

[14] The limit distribution of this difference can easily be obtained under the null hypothesis.

3.6 Conclusions

In this chapter we try to explain the reasons why contracting out refuse collection is less common than in-house provision, although considerable efficiency improvements by contracting out seem achievable. We present an empirical investigation motivated by output arguments, interest group theory, and ideology arguments.

We used both a parametric (logit) and a semiparametric (Ichimura, 1993) modeling approach, which correspond in the use of a single index, but which differ in terms of the flexibility of the probability transformations employed. The estimated single indices are quite similar, so that both yield the same conclusions, when investigating the direction and statistical significance of the various effects.

In both models we find evidence for the hypothesis that a high level of transfers by the central government (the wealth argument) or a high level of unemployment (the interest group argument) raises the probability of public and in-house provision. We also find evidence for the assumed relation between the size of municipalities and private collection. In all cases a smaller municipality is more likely to have private collection. Therefore, scale effects are important for the choice between public and private provision. For the choice between out-house and in-house collection in relation to scale lesser evidence exists. Weak evidence is found for an ideological motivation of this choice.

However, when explicitly quantifying the size of these effects, one also needs the probability transformation, transforming the single index into the probability that the dependent variables equals one. Here, we find strong statistical evidence that the parametric specification is far too inflexible, with the danger that the corresponding estimated marginal effects might be misleading. Indeed, in a number of cases, we find serious differences between the parametric and the semiparametric marginal effects, implying that one should be very cautious, when using parametric models.

Acknowledgments We thank two anonymous referees for their comments on an earlier version. This chapter was published as Dijkgraaf E, Gradus RHJM, Melenberg B (2003), Contracting out refuse collection. Empirical Economics 28: 553-570.

References List

Buchanan, J. (1987). The constitution of economic policy. *American Economic Review*, 77, 243–50.

Bivand, R., & Szymanski, S. (2000). Modelling the spatial competition of the introduction of Compulsory Competitive Tendering. *Regional Science and Urban Economics*, 30, 203–219.

Bosch, N., Pedaja, F., & Suarez-Pandiello, J. (2000). Measuring the efficiency of Spanish municipal refuse collection services. *Local Government Studies*, 26, 71–90.

Dijkgraaf, E., & Gradus, R. H. J. M. (1997). *Cost savings of contracting out refuse collection* (Research Memorandum 9712). Erasmus University Rotterdam (published as Dijkgraaf E, Gradus RHJM (2003). Cost savings of contracting out refuse collection. *Empirica*, 30, 149–161 and reproduced in chapter 2).

Domberger, S., & Jensen, P. (1997). Contracting out by the public sector: theory, evidence and prospects. *Oxford Review of Economic Policy*, 4, 67–78.

Domberger, S., Meadowcroft, S., & Thompson, D. (1986). Competitive tendering and efficiency: the case of refuse collection. *Fiscal Studies*, 7, 69–87.

Gourieroux, C., Laffont, J., Montfort, A. (1980). Coherency conditions in simultaneous linear equation models with endogenous switching regimes. *Econometrica*, 48, 675–695.

Hart, O., Shleifer, A., & Vishny, R. (1997, November). The proper scope of government: theory and applications to prisons. *Quarterly Journal of Economics*, 1127–1161.

Horowitz, J. (1992). Semiparametric estimation of a work-trip mode choice model. *Journal of Econometrics*, 58, 49–70.

Horowitz, J. (1998). Semiparametric methods in econometrics. Lecture Notes in Statistics (Vol. 131). New York: Springer.

Ichimura, H. (1993). Semiparametric least squares (SLS) and weighted SLS estimation of single-index models methods in econometrics. *Journal of Econometrics,* 58, 71–120.

Kitchen, H. M. (1976). A statistical estimation of an operating cost function for municipal refuse collection. *Public Finance Quarterly*, 4, 43–72.

Klein, R. W., & Spady, R. H. (1993). An Efficient Semiparametric Estimator for Binary Response Models. *Econometrica*, 61, 387–421.

López-de-Silanes, F., Shleifer, A., & Vishny, R. W. (1997). Privatization in the United States. *Rand Journal of Economics*, 28, 447–471.

Manski, C. F. (1993). Indentification of endogenous social effects: The reflection problem. *Review of Economic Studies*, 60, 531–542.

Manski, C. F. (1985). Semiparametric analysis of discrete response: Asymptotic properties of the maximum score estimator. *Journal of Econometrics*, 27, 313–333.

Newey, W. K. (1985). Maximum Likelihood Specification Testing and Conditional Moment Tests. *Econometrica*, 53, 1047–1070.

NG magazine (1998). Uitbesteding van afvalinzameling niet altijd beter. 20 November, 16–17.

Ohlsson, H. (1998). *Ownership and production costs: Choosing between public production and contracting out.* Uppsala University working paper 6.

Ruud, P.A. (1984). Tests of specification in econometrics. *Econometric Reviews*, 3, 211–242.

Reeves, E., & Barrow, M. (2000). The impact of contracting out on the costs of refuse collection services: The case of Ireland. *Economic and Social Review*, 31, 129–150.

Schmidt, P. (1981). Constraints in simultaneous Tobit and Probit models. In F. Manski, D. McFadden (Eds.), *Structural analysis of discrete data with econometric applications* (pp. 422–434). Cambridge Mass: MIT Press.

Stevens, B. J. (1978). Scale, market structure and the cost of refuse collection. *Review of Economics and Statistics*, 60, 438–448.

Szymanski, S., & Wilkins, S. (1993). Cheap rubbish? Competitive tendering and contracting out in refuse collection – 1981–1988. *Fiscal Studies*, 14, 109–130.

Szymanski, S. (1996). The impact of compulsory competitive tendering on refuse collection services. *Fiscal Studies*, 17, 1–19.

Tang, K. L. (1997). Efficiency of the private sector: a critical review of empirical evidence from public services. *International Review of Administrative Sciences*, 63, 459–474.

Chapter 4
Contracting Out in Sweden: Ownership and Production Costs

H. Ohlsson

Abstract Many comparisons of public and private firms use a public/private ownership dummy variable to capture cost differences. If, however, public and private firms use different production technologies, the dummy-variable approach is misspecified. Data from public and private firms should not be pooled. Secondly, selectivity bias may arise, making it more difficult to identify cost differentials that actually exist. Thirdly, if data should be pooled, the resulting empirical model may be logically inconsistent. This chapter compares public and private firms using the refuse collection costs of 170 firms in 115 Swedish municipalities. First, public production costs were 6 per cent lower than private production costs. Secondly, cost differences did not affect producer choice. It is crucial to adjust for selectivity. Data for private and public firms should not be pooled. The dummy-variable model is misspecified.

Keywords Contracting-out · ownership · collection · Sweden · privatization

4.1 Introduction

The public sector has expanded in all industrialized countries and the tax pressure has increased. These trends, together with ideological changes, have meant that public activities during the last few decades have been questioned more and more. There have been extensive privatizations in many countries in response to this.[1] Public ownership and the efficiency of public firms have been widely discussed.

There are, however, considerable methodological problems when studying the effects of ownership. The discussion about how to compare the performance of private and public firms clearly illustrates the difficulties in measuring firm performance.

✉ H. Ohlsson
Department of Economics, Uppsala University, P.O. Box 513, SE-751 20 Uppsala, Sweden
e-mail: henry.ohlsson@nek.uu.se

[1] The literature on privatization is extensive; Vickers and Yarrow (1988) and Bös (1991) are two early contributions.

The market structure and the degree of *competition* is one issue. The performance of public and private firms may not differ in a competitive environment. Without competition, however, firms will have the opportunity to extract monopoly rents.

Then the *objectives* of firms become crucial. Private and public firms may differ in objectives. Public firms may not exploit their market position; instead, they may, for example, maximize social welfare. Profit-maximizing private firms, on the other hand, will exploit their market position if there is not enough competition.

Private and public firms may also differ in *corporate governance*. This may concern how managers are monitored or the incentive structure of managers. Principal-agent considerations show that there are more complexities comparing firm performance. It is often conjectured that public firms are less efficient internally because public managers put less effort into reducing costs than private managers.

At the same time, some argue that the default risk of private firms results in higher capital costs than for public firms. By the same argument, it is conjectured that private firms are also more likely to pay more for inputs.

Costs and profits are the evaluation criteria most often used when empirically comparing the performance of private and public firms. However, profits measure efficiency poorly when there is lack of competition. This is an argument for using costs instead. Pestieau and Tulkens (1993) argue that it is impossible to sort out the ownership matter from the market structure and regulation matters. Instead, they advocate that the performance of public firms should be measured and compared on the basis of productive efficiency only.[2]

There is a considerable empirical literature on these issues. Vining and Boardman (1992) survey the empirical literature on the effect of ownership on efficiency.[3] The survey includes, for example, previous studies on the costs of refuse collection. Vining and Boardman argue that ownership in itself has a role separate from the degree of competition in the output market. The position of Borcherding, Pommerehne, and Schneider (1982), on the other hand, is that there are no ownership effects when controlling for competition. There are also studies focusing on contracting out in particular. Domberger and Jensen (1997) conclude that the evidence shows that contracting out may reduce costs considerably.

Many studies focus on the possible differences in costs between private and public firms. On the other hand, few focus on what determines the political decision-makers' choice between private and public firms. Understanding this choice is as important as understanding possible cost differences.

The starting point for this chapter is that many previous studies of the performance of public and private firms use dummy variables to capture the effects of public/private ownership (or market organization in general) in cross-sections.[4] When estimating cost functions, the dummy variables are intended to capture the

---

[2] More specifically, they argue for using a frontier approach for example, data envelopment analysis (DEA) for measuring productive efficiency.

[3] Other surveys are Bennett and Johnson (1980), Pestieau and Tulkens (1993) and Tang (1997).

[4] Szymanski and Wilkins (1993) and Szymanski (1996) are exceptions. These studies use panel data.

cost difference between, for example, public and private firms. The model has, however, nothing to say about producer choice. In addition, this approach raises three important empirical issues: pooling, selectivity and logical consistency.

- *Pooling.* If public and private firms use different production technologies, the dummy-variable approach is simply misspecified. Data from the different types of firms should not be pooled if production technologies differ.
- *Selectivity.* A second potential problem arises because the type of producer is chosen by the public authorities. The choice is not random.[5] There is, therefore, a risk of selectivity bias when estimating cost functions. A consequence of this is that the cost differences may be overestimated. This will be a potential problem regardless of whether the dummy-variable model or some other specification is used.
- *Logical consistency.* Thirdly, the dummy-variable approach is logically inconsistent if private firms only have lower costs in the cases when private production is chosen.[6] If, on the other hand, private firms always produce at lower costs, why do we observe public production at all if costs matter for the choice of producer?[7]

The dummy-variable approach may be appropriate in an experimental situation, when the type of producer is chosen randomly. To my knowledge, there are, however, no data available from experiments with random assignments of the type of producer. Instead, an empirical cost comparison fits well in the econometric framework of a switching regression model with endogenous switching. It takes producer choice into account and provides a unified framework to test pooling, selectivity and logical consistency. The objective of this chapter is to estimate a model of this kind using Swedish data.

In the case here, it turns out that it is crucial to adjust for selectivity and that it is not correct to pool data for private and public firms. The dummy-variable model is misspecified. As data should not be pooled in this case, the problem of logical consistency does not arise. The two main results are as follows:

- The estimations suggest that public production, on average, was 6 per cent cheaper than private production.
- Cost differences did not affect producer choice. The municipalities, in other words, did not choose the least-cost alternative.

Model selection is crucial. Suppose that we erroneously pool the data and do not correct for selectivity by using the dummy-variable model. This gives the opposite qualitative result: costs in public firms are estimated to be 13 per cent higher than those in private firms. If, instead, we correct for selectivity but still erroneously

[5] Marcinèin and van Wijnbergen (1997) show that it is crucial to take selection into account also when studying the effects of privatization.

[6] The terminology is from Maddala (1983). Heckman (1978) and Maddala and Lee (1976) also discuss this.

[7] Dubin and Navarro (1988, 219) ask: 'if private monopoly is the efficient, cost-minimizing alternative, why do over half of the local communities choose other forms of market organization?'

pool the data, the public cost advantage is overestimated: costs in public firms are estimated to be 36 per cent lower than those in private firms.

The chapter is organized as follows. The data and the decision-making process are presented in Sections 4.2 and 4.3 discusses the potential problems with the dummy-variable approach. It also specifies the more general model and discusses how it should by estimated. Section 4.4 presents a theoretical framework for the choice of producer by local public authorities. The estimations of producer choice models and cost models can be found in Sections 4.5 and 4.6 concludes.

4.2 Data and the Decision-Making Process

The data-set comes from a survey carried out by the Swedish Competition Authority in 1989. It covers refuse collection in 115 of Sweden's 284 municipalities. There were two major municipal boundary reforms in Sweden during the 1950s and 1960s. The number of municipalities was reduced from 2,500 in 1952 to 278 in 1974. Since then, there has been some breaking-up of municipalities.

Since 1972, the Swedish municipalities have had local monopolies by law in the provision of refuse collection.[8] A municipality can choose to collect itself or it can decide to contract out the collection. In 1989, when the data were collected, procurement in almost all municipalities was made according to procurement guidelines suggested by the Swedish Association of Local Authorities. These guidelines stressed that the municipalities should take advantage of existing possibilities for competition when procuring. The guidelines also stressed that procurement should be done in a businesslike manner and that all bids should be treated objectively. There was a recommendation to have competitive tendering using sealed bids. This type of procurement was, however, not compulsory.[9]

A municipality can decide to procure refuse collection from different firms in different areas of the municipality. (Sometimes, the collection areas correspond to the 'old', small, pre-amalgamation municipalities.) Several private firms may collect in different areas of the municipality. It may also be the case that the municipality collects itself in some areas while there is contracting out in other areas. There is, however, only one collector in each collection area.

In 56 municipalities, the collection was, completely (35) or partly (21), done by the municipality. In the remaining 59 municipalities, private firms were the only collectors. All in all, 150 'firms' were involved, 55 public and 95 private. Some

---

[8] Why should refuse collection be compulsory and publicly provided? What is the market failure? Consumption of collection services is rival and exclusion is possible. The reasons are general health and sanitation. Externalities exist because individuals are jointly damaged by deteriorations in the environment when some individuals choose low (or no) levels of collection services. The deteriorations are characterised by indivisibilities and exclusion is difficult or impossible.

[9] Currently, public procurement must be done in accordance with the 1992 public procurement law that came into effect in 1994. This law is much stricter than the previous guidelines.

Table 4.1 Types of procurement and number of collection areas

	Private firm chosen	Public firm chosen
Competitive tendering, sealed bids	23	10
Other types of procurement	77	43
Missing	14	3

firms, however, operated in more than one municipality. The data for these firms are separated for each municipality so the number of firms at the municipality level is 170, 56 public and 114 private. Appendix 1 presents some more information about the variables used. There are unfortunately many missing observations for some variables, particularly the cost variables.[10]

Table 4.1 reports the type of procurement that was used. It is clear from the table that competitive tendering with sealed bids was only used in one in five cases.

There was little difference in the proportionate use of sealed-bid competitive tendering whether private or public firms were chosen. The most important other types of procurement were direct procurement and bargained procurement.

There is some information available on the number of bids, but unfortunately only in the cases when private firms were chosen. When procurement was done using sealed-bids competitive tendering, there were two to six bids in most cases. Information on the number of bids is available for all 23 cases of sealed-bids competitive tendering when a private firm was chosen. There was usually only one bid when the other types of procurement were used. There are, however, many missing observations for these cases.

In some cases, the municipalities claimed that there were no private firms interested in providing the service. One way of increasing the competitive pressure when the number of firms is not large enough is to analyze the collection costs. Some procuring municipalities have, therefore, done a thorough analysis of the collection costs in particular collection areas. The objective is to create yardstick competition. The information collected in this way allows the municipalities to identify the least-cost alternative more easily. Table 4.2 shows that such an analysis was carried out in 52 of the cases when competitive tendering with sealed bids was not used. A cost analysis also complemented competitive tendering with sealed bids in many cases.

Table 4.2 Cost analysis and number of collection areas

	Cost analysis done	No cost analysis done
Competitive tendering, sealed bids	22	11
Other types of procurement	52	68

[10] In most estimations of cost functions, I can only use 47 public firms and 75 private firms. As far as I can test, however, this subsample is representative of the full sample.

4.3 Models and Estimation Strategy

4.3.1 How it has Usually Been Done

The most common empirical approach used when comparing public and private firm performance is to estimate a single cost function with production quantity and factor prices as explanatory variables and simply adding a dummy variable for the type of ownership. In (SPK Swedish National Price and Cartel Office (1991)), the official government report using the data-set I use, the total costs of private refuse collection were estimated to be 25 per cent lower than the costs of public production.[11] Total costs include costs for labor, vehicles, offices, packing and some minor items. Re-estimating the equation, adding housing density as an explanatory variable in one specification, I get the results reported in Table 4.3. [12]

There are four variables measuring different dimensions of produced output. These are the quantity collected, the number of places from which refuse has been picked up in relation to the quantity collected, the number of pick-ups in relation to the number of pick-up points, and the distances driven. Previous refuse collection studies have also used similar output variables.

All estimated parameters for the output variables are significant at the 5 per cent level except for pick-up frequency. Factor prices are borderline significant. I have also added housing density as this variable has been included in many previous studies. It is significant. Costs in private firms are 12–20 per cent lower than costs in public firms, depending on which other variables are included.[13] The reason I cannot replicate the estimations in the government report is that the agency chose to impute many missing values whereas I do not use these imputed values. The private-ownership dummy is significant in the estimation reported in column 3.

[11] There are many papers studying ownership effects on refuse collection costs using the dummy-variable approach. Among them are Hirsch (1965), Kemper and Quigley (1976), Kitchen (1976), Collins and Downes (1977), Pommerehne and Frey (1977), Stevens (1978), McDavid (1985), Domberger, Meadowcroft, and Thompson (1986), Tickner and McDavid (1986), Dubin and Navarro (1988), Szymanski and Wilkins (1993), Szymanski (1996), Gradus and Dijkgraaf (1997) and Reeves and Barrow (2000). Edwards and Stevens (1978) use the dummy-variable method to study refuse collection prices. There are also other studies of refuse collection. Cubbin, Domberger, and Meadowcroft (1987) and Bosch, Pedraja, and Suàrez-Pandiello (2000) use DEA approaches to study productive efficiency. Some studies simply compare refuse collection costs or prices without controlling for other variables. Savas (1977a, 1981) and Bennett and Johnson (1979) are among these

[12] I have used the LIMDEP version 7.0 software package (see Greene, 1995) and the Stata release 7.0 software package (see StataCorp, 2001).

[13] Most of the previous studies using the dummy-variable method also find that private production is cheaper. Collins and Downes (1977), however, report the opposite result.

Table 4.3 Costs per tonne: dummy-variable models

	(1)	(2)	(3)
Quantity	−0.11	−0.14	−0.22
	(1.91)	(2.63)	(3.72)
Pick-up points per tonne	0.20	0.21	0.15
	(2.83)	(3.30)	(2.46)
Pick-up frequency	0.02	−0.02	−0.06
	(0.19)	(0.27)	(0.68)
Distance	0.15	0.13	0.14
	(3.17)	(2.86)	(3.10)
Wage rate		0.23	0.18
		(1.90)	(1.55)
Cost of capital		0.18	0.21
		(1.78)	(2.14)
Housing density			0.09
			(2.93)
Private ownership	−0.13	−0.12	−0.20
	(1.28)	(1.25)	(2.10)
Constant	−0.21	0.74	1.38
	(0.55)	(1.60)	(2.78)
R^2	0.17	0.25	0.31
SEE	0.48	0.43	0.41
RSS	27.19	19.78	18.31
Log likelihood	−81.54	−62.00	−57.53
Number of observations	122	116	116

Notes: All variables are in logarithms, including the dependent variable, except the private-ownership dummy. Absolute t-values are given in parentheses

4.3.2 Why it May be Wrong

There are several potential problems related to empirical estimations. One is misspecification. Is it correct to pool data from private and public firms? Suppose that the cost functions are

$$\text{Private costs: } C_{pr} = \beta_{pr} X + u_{pr} \qquad (4.1a)$$

$$\text{Public costs: } C_{pu} = \beta_{pu} X + u_{pu} \qquad (4.1b)$$

where C_{pr} and C_{pu} are costs, the vector X includes production quantity and factor prices, while u_{pr} and u_{pu} are error terms. The subscripts pr and pu refer to private production and public production.[14] Suppose that the cost functions are the same for all firms ($\beta_{pr} = \beta_{pu}$ for all X). We may then pool the data and estimate a single cost function. Having the same production technologies implies that the cost functions

[14] Not all explanatory variables may have an influence in both equations. For some elements of X, the β_{pr} and β_{pu} coefficients in the cost functions may be zero.

are the same. But if the production technologies differ, we must allow public and private firms to have different cost functions (different βs).

A second problem is selectivity. This may arise if the producer choice is not completely random. Suppose that producer choice is determined according to

$$\text{Producer choice: } Pr^{*} = \gamma Z + \delta \left(C_{pr} - C_{pu}\right) + v \tag{4.2}$$

where Pr^{*} is a latent variable for private firms with the corresponding binary variable Pr,[15] Z is a vector of variables that influence producer choice and v is an error term. The term $C_{pr} - C_{pu}$ captures the cost difference between private and public production. Some, but not all, of the variables in X may also appear in Z.

We will also have use for the reduced form of the choice equation. Here, the cost difference is replaced by the determinants of costs according to (4.1a) and (4.1b). The reduced form is

$$Pr^{*} = \gamma Z + \delta \left(\beta_{pr} - \beta_{pu}\right) X + \delta \left(u_{pr} - u_{pu}\right) + v \tag{4.3}$$

which can be given new parameters to become $Pr^{*} = \tilde{\gamma} Z + \tilde{v}$.

The potential selectivity problem can be illustrated by the following example. Suppose that firms differ in production costs partly because of differences in, for example, managerial ability. Furthermore, let us assume that there is no information available to us about this ability. Now suppose that we study a firm with higher managerial ability, and therefore lower production costs, than captured by the exogenous variables in the cost equation. This will give rise to a negative error term u_{pr} if it is a private firm and a negative error term u_{pu} if it is a public firm.

The procuring municipalities may, however, have learnt over time about the managerial ability of different firms. Suppose that this reputation about ability makes the municipality more likely to procure from high-ability firms than captured by the explanatory variables in the producer choice equation. We will then have a positive error term v if it is a private firm (and a negative error term v if it is a public firm). More importantly, there will exist a negative covariance, $\sigma_{u_{pr} v}$ between the error term in the private cost equation, u_{pr}, and the error term in the choice equation, v. There will also exist a positive covariance, $\sigma_{u_{pr} v}$, between the error term in the public cost equation, u_{pu}, and the error term in the choice equation, v. Because of reputation, high-ability firms are more likely to be chosen.

Suppose that these covariances are indeed non-zero. In general, $\sigma_{u_{pr} v}$ can then be expected to be negative while $\sigma_{u_{pr} v}$ can be expected to be positive. If we now estimate the cost equations without taking the covariances into account, the estimation results will suffer from selectivity bias. Producer choice will give us more observations of firms with true cost error terms being negative than observations of firms with positive error terms. But empirical models that do not control for selectivity assume that there are as many positive error terms as negative. The estimated coefficients will, therefore, be biased.

---

[15] The observed variable $Pr = 1\, if\, Pr^{*} > 0$ and $Pr = 0$ if $Pr^{*} \leq 0$.

The conditional expected costs are

$$E\left(C_{pr}\,|Pr = 1\right) = \beta_{pr}X + \sigma_{u_{pr}v}\frac{\phi(\tilde{\gamma}\tilde{Z})}{\Phi(\tilde{\gamma}\tilde{Z})} \tag{4.4a}$$

$$E\left(C_{pu}\,|Pr = 0\right) = \beta_{pu}X + \sigma_{u_{pu}v}\frac{\phi(\tilde{\gamma}\tilde{Z})}{1-\Phi(\tilde{\gamma}\tilde{Z})} \tag{4.4b}$$

where $\phi(\tilde{\gamma}\tilde{Z})$ and $\Phi(\tilde{\gamma}\tilde{Z})$ are the density function and the distribution function of the standard normal evaluated at $\tilde{\gamma}\tilde{Z}$. Including $+\sigma_{u_{pr}v}(\phi/\Phi)$ or $-\sigma_{u_{pu}v}(\phi/(1-\Phi))$ when estimating the respective cost equation will control for selectivity and yield estimates of the covariances.

If we estimate the cost equations (4.1a) and (4.1b) without controlling for selectivity, we will get biased estimates if the covariances are non-zero. Most probably, we will tend to overestimate the cost advantage of private ownership. This will also, in a second step, affect the estimation of the choice equation. Suppose that we have biased estimates of the coefficients of the cost equations and therefore a biased estimate of the cost difference. This will lead to biased estimates of the producer choice equation regardless of whether we estimate (4.2) using the expected cost difference as explanatory variable or (4.3) using the differences in estimated coefficients times the X variables as explanatory variables.

A third potential problem is logical inconsistency. This problem may arise if we pool the data. Suppose that we capture the cost effect by a dummy variable when estimating the cost equation. The issue now is whether this effect should be attributed to the type of firm ownership or to the particular area where the refuse is collected. If, on the one hand, it is connected to the area, all firms will have to bear these costs, which are the same for all. In this case, there will be no variation in costs across firms. Consequently, it is not possible to identify the impact of ownership when estimating a choice equation.

If, on the other hand, the cost difference should be attributed to ownership, there will be variation across firms. Private firms will have different costs from public firms. The model is

$$\text{Costs: } C = \beta X + \alpha\,\text{Pr} + u \tag{4.5a}$$

$$\text{Producer choice:} Pr^{*} = \gamma Z + \delta\alpha\,\text{Pr} + v \tag{4.5b}$$

where α is the ownership effect. The parameter α <0 if private production is cheaper. However, Maddala (1983, 118) presents the following lemma (I use the notation of (5a) and (5b)):[16]

Lemma 1. *Suppose Pr^* is an unobserved variable, with the corresponding observed variable $Pr = 1$ if $Pr^* > 0$ and $Pr = 0$ if $Pr^* \leq 0$. Then a model of the form $Pr^* = \gamma Z + \delta\alpha Pr + v$, where Z is a variable and γ is a parameter, is logically inconsistent unless $\delta\alpha = 0$.*

[16] Heckman (1978, 936) also provides a proof of this proposition.

Proof: $P(Pr = 0) = 1 - F(\gamma Z)$ while $P(Pr = 1) = F(\gamma Z + \delta\alpha)$ where $F(\cdot)$ is the distribution function. The probabilities sum to one, i.e. $1 - F(\gamma Z) + F(\gamma Z + \delta\alpha) = 1$. But this holds only if $\delta\alpha = 0$.

Logical consistency requires that the cost difference plays no role for the producer choice, i.e. $\delta = 0$. Alternatively, there should be no cost difference, i.e. $\alpha = 0$. Cost differences will then play no role in the choice to contract out. To simplify matters, suppose also that there are no other variables affecting choice ($\gamma Z = 0$). The producer choice, or rather the assignment, is random according to υ.

4.3.3 How it Could be Done

The combination of the cost equations (4.1a) and (4.1b) and the producer choice equation (4.2) is a switching regression model with endogenous switching. We can use it to estimate whether there are cost differences between private and public firms and whether cost differences affect producer choice. At the same time, the model provides a unified framework for testing selectivity, pooling and logical consistency. Some of the previous empirical studies of refuse collection have done parts of what could be done. But no study has done it all at the same time.

How do I test selectivity, pooling and logical consistency? If the error covariances $\hat{\sigma}_{u_{pr}\upsilon}$ and $\hat{\sigma}_{u_{pu}\upsilon}$ are significant in the cost equations, we know that this correction for sample selection was indeed needed.

I test pooling in the following way. Maddala (1983) suggests an empirical specification where the expected cost is

$$E(C) = \beta_{pu}X + \left(\beta_{pr} - \beta_{pu}\right) X\Phi(\gamma Z) + \left(\sigma_{u_{pr}\upsilon} - \sigma_{u_{pu}\upsilon}\right) \phi(\gamma Z) \quad (4.6)$$

The cost equation (4.6) can be estimated using a two-step procedure. Suppose that $\beta_{pr} = \beta_{pu}$ for all X except the constant. This model is sometimes called the treatment effects model. It can be viewed as a restricted version of the selectivity-controlled cost equations, in the sense that all the coefficients, except the constants, are the same. Equation (4.6) then collapses to

$$E(C) = \beta X + \alpha\Phi(\gamma Z) + \left(\sigma_{u_{pr}\upsilon} - \sigma_{u_{pu}\upsilon}\right) \phi(\gamma Z) \quad (4.7)$$

where α equals the difference in the β coefficients for the constant. This equation captures the effect of ownership on costs. Pooling for the whole model can be tested by a likelihood ratio test of the treatment effects model versus the separate cost equations. If pooling for the whole model is rejected, we can then test pooling variable by variable by estimating (4.6). The impact of a variable will differ between private firms and public firms if the relevant estimated coefficient in $\beta_{pr} - \beta_{pu}$ is significant.

There will be no problem of logical inconsistency if pooling is rejected. If data should be pooled, consistency requires either that there is no cost difference ($\alpha = 0$) or that the cost difference does not affect producer choice ($\delta = 0$). In the pooling case, the estimation of the treatment effects model provides a test of $\alpha = 0$, while the structural form probit gives the test of $\delta = 0$.

4.4 Theoretical Framework

Which variables form X and Z? Cost functions usually follow directly from the assumption that the firm optimizes. Costs will then be a function of quantity produced and factor prices. The effects of factor prices can be identified empirically if factor prices vary between different collection areas. We can then compare public and private costs for a given quantity collected and given, exogenous, area-specific factor prices.

Suppose instead that factor prices vary between public and private firms. The comparison can then only be made for a given quantity, as factor prices are endogenous (ownership-specific). There are systematic differences in the factor prices paid by public and private firms in the data-set I use. In a previous study using the same survey (Ohlsson, 1996), I found that private firms, controlling for other factors, pay 10–15 per cent less for their trucks. As firms cannot be assumed to be price-takers, factor prices should not be included in the cost functions. I will therefore assume that the X vector only consists of variables capturing different dimensions of the quantity produced.

Gómez-Lobo and Szymanski (2001) find that more bids in compulsory competitive tendering are associated with lower costs for refuse collection. Here, it is not possible to test this when public firms are chosen because of lack of data. When testing whether the number of bids affects costs, I have therefore had to restrict the estimation to the cases when private firms have been chosen. I cannot, however, find any significant effects of the number of bids.[17]

4.4.1 Transaction Costs

The production costs in private firms are not necessarily the same as the payment of the public sector when choosing to contract out. I will, however, assume that cost-plus contracts are used. I will also assume that there are no differences in the possibilities for the public sector to forecast the costs of own production and the costs when contracting out.

These are, of course, strong assumptions. Real resources will have to be spent both during and after the actual procurement. Borcherding (1988) and Wittman (1989) argue that, on one hand, observed cost differences can be attributed to cost-increasing behavior of public managers. Many empirical studies, on the other hand, implicitly assume that the transaction costs associated with contracting out are

17 The results are available from the author on request.

negligible. When total costs are considered, this may make public production more efficient.

There are also transaction costs associated with bargaining, monitoring and enforcing when contracting out. The public firms, which are perceived to be inefficient, may actually be the least-cost alternative when all things are considered. Contracting costs and contract failure are also discussed by Domberger and Jensen (1997).

4.4.2 Producer Choice

Things are less clear for how (local) politicians choose the producer of publicly provided goods and services. De Silanes, Schleifer, and Vishny (1997) discuss three types of determinants of producer choice: efficiency (social goals), political patronage and ideology. Political patronage has to do with the fact that politicians get support from public employees when services are publicly produced. Politicians will therefore favor public production unless taxpayers force them to do otherwise. Hoover and Peoples (2003) find that US municipalities are more likely to use union refuse workers when a relatively large percentage of the residents in the municipality are union members.

Similar determinants have been suggested by other authors. Ferris (1986) writes that contracting out is more likely with greater cost savings. More stringent fiscal limits and less powerful public employees and public service constituency groups will also make contracting out more likely. Nelson (1997) adds that heterogeneous citizen preferences may make contracting out less likely.[18]

To sum up, four different types of determinants of producer choice have been suggested in the literature:

- ideology;
- cost differences (efficiency);
- the influence of pressure groups (taxpayers, public employees, etc.);
- the pressure from legal constraints on fiscal behavior.

4.5 Evidence: Costs and Producer Choice

4.5.1 Representative Sample

There are many missing observations for the firm-level variables, while the municipality-level data I use in the choice models are complete. There is, therefore, a potential problem that the sample is not representative. Table 4.4 reports some probit estimations that address this problem.

Column 1 of the table is an ownership probit for a sample of 153 firms. Information on whether the municipality has used sealed-bids competitive tendering

---

[18] The determinants of producer choice are also discussed in Sonenblum, Kirlin, and Ries (1977), Dubin and Navarro (1988), Szymanski (1996) and Dijkgraaf, Gradus, and Melenberg (2003).

Table 4.4 Producer choice and test of representative sample: probit models

	Dependent variable		
	Private firm =1	Private firm =1	Cost data available =1
	Sample		
	All firms (1)	Firms for which cost data are available (2)	All firms (3)
Socialist	0.01	0.22	0.30
	(0.06)	(0.77)	(1.10)
Share of single-family houses	2.51	2.86	-0.02
	(1.98)	(1.95)	(0.02)
Competitive tendering, sealed bids	0.24	0.41	0.16
	(0.84)	(1.33)	(0.54)
Cost analysis made by the procuring municipality	-0.54	-0.64	-0.15
	(2.32)	(2.46)	(0.59)
Average income	0.15	0.09	-1.85
	(0.07)	(0.03)	(0.92)
Population	-0.36	-0.47	0.43
	(0.68)	(0.69)	(1.35)
Population density	1.35	2.68	7.03
	(0.23)	(0.38)	(1.08)
Housing density	-0.29	-2.14	-13.5
	(0.03)	(0.16)	(1.13)
Constant	-0.93	-1.28	2.24
	(0.41)	(0.48)	(1.11)
Log likelihood	-88.0	-70.8	-75.0
Average likelihood	0.56	0.56	0.61
χ^2	21.5	21.1	4.13
Significance level	0.006	0.007	0.845
Number of observations	153	122	153

Note: Absolute t-values are given in parentheses.

and/or on whether the municipality has done a cost analysis of its refuse collection is missing for 17 firms, so the sample size in column 1 is reduced from the full sample of 170 firms. Column 2 is a corresponding probit for the sample of 122 firms for which cost data are available. The estimated coefficients do not differ considerably between the two estimations.

Running a probit on whether cost data are available (column 3) does not give any particularly significant coefficients. My conclusion is that the subsample with available cost data is not biased. A χ^2 test of the restriction that the model only has a constant does not reject the restriction.

4.5.2 Producer Choice

Unfortunately, there are few variables available that directly correspond to the determinants of producer choice discussed in Section 4.4. However, χ^2 tests of the restriction that the models in Table 4.4, columns 1 and 2, only have a constant reject the restriction in both cases. In this sense, the estimated models as such have explanatory power.

I have proxied ideology with a dummy variable for if the Social Democrats and the Left Party together had a majority in the municipality council. This dummy variable is, however, not significant.[19] It could be argued that the share of single-family houses in the housing stock could proxy ideology. The idea is that there is a correlation between preferences for owning a single-family house and more conservative political preferences. This variable has a significantly positive impact on the probability that a private firm is chosen.

Efficiency considerations are captured by two dummy variables. Using competitive tendering with sealed bids seems to make choice of a private firm more likely. The estimated coefficients are, however, far from significant. Some procuring municipalities have analyzed the collection costs in the particular collection areas to create yardstick competition. The estimated coefficient on the dummy variable indicating this is significant: cost analyses increase the probability that the procuring municipality chooses public production. This is the most important determinant of producer choice together with the share of single-family houses.

I also have access to the variables average income, population, population density and housing density. These variables may measure the influence of pressure groups. However, none of these variables does have a significant impact on the probability of the municipality choosing a private firm.[20]

Dubin and Navarro (1988) actually estimate choice equations (Szymanski (1996) discusses the matter). They find that the probability of public ownership is lower if the share of votes for the Democratic Party is high and if the fraction of unionized refuse collection workers is high, while it is decreasing in per capita income.

4.5.3 Cost Functions

Table 4.5 presents estimations of cost functions controlling for sample selection. The dependent variable is the logarithm of costs per tonne. The probit in Table 4.4, column 3, is used to compute the sample-selection-term variables, which equal ϕ/Φ for private firm costs and $-\phi/(1-\Phi)$ for public firm costs. Davidson and MacKinnon (1993) write that two-stage least squares (2SLS) should be used to test for selectivity bias while maximum likelihood estimation should be used if selectivity bias cannot be rejected. Table B1 in Appendix 2 presents maximum likelihood estimations. The results are, in general, similar to those reported in Table 4.5.

---

[19] Bivand and Szymanski (2000) and Gómez-Lobo and Szymanski (2001), on the other hand, find political effects on costs using UK data. Local governments led by the Conservatives have lower costs than those led by Labour.

[20] I have tried several other variables capturing different aspects of the finances of the municipality without getting any significant result. Among the variables tried without success are expenditure per capita, revenue per capita, central government grants per capita, tax base per capita, tax rate, tax revenue per capita, net worth per capita and the share of the population reporting income.

Table 4.5 Cost functions and tests for selectivity and pooling: sample selection models

	Private firms (1)	Public firms (2)	All firms (3)
Quantity	-0.13	-0.06	-0.09
	(1.65)	(1.23)	(1.64)
Pick-up points per tonne	0.16	0.45	0.20
	(1.78)	(5.49)	(2.88)
Pick-up frequency	-0.01	0.22	0.01
	(0.05)	(1.49)	(0.10)
Distance	0.21	0.05	0.16
	(3.09)	(1.46)	(3.41)
Private ownership			0.36
			(1.38)
Selection term	-0.45	0.04	-0.34
	(1.82)	(0.33)	(2.11)
Constant	0.07	-1.06	-0.50
	(0.14)	(1.99)	(1.24)
σ_{u_i}	0.55	0.23	0.46
ρ	-0.71	0.18	-0.64
R^2	0.20	0.52	0.21
RSS	20.50	2.08	24.53
Log likelihood	-57.78	6.57	-75.26
Number of observations	75	47	122

Notes: All output variables are in logarithms, including the dependent variable costs per tonne. The models are estimated using 2SLS. Absolute t-values are given in parentheses.

The sample selection term is borderline significant in the private cost function. It also has the predicted sign. The sample selection term is significant in the maximum likelihood estimation reported in Appendix 2. The public cost function, on the other hand, seems to be unaffected by the selection term.

This implies that there is a negative and significant covariance between the error term in the producer choice equation and the error term in the private firm cost function. Private firms that are chosen by municipalities, although they are not predicted to be chosen, have lower costs, controlling for the other variables in the cost function.[21]

Distance is the output variable with the highest t-statistic in the private cost function, while pick-up points per tonne is borderline significant (Table 4.5, column 1).[22] In the public cost function, pick-up points per tonne is the only significant output variable (column 2). These differences raise the question of whether it is correct to pool data from private and public firms.

[21] Dubin and Navarro (1988) also test for selectivity, but this is rejected in their case.

[22] I have tried the dummy variables for sealed-bids competitive tendering and cost analysis in the cost functions. The estimated coefficients were not significant. I have also tried the number of bids in the cost function for private firms. The estimated coefficient was not significant. The estimation results are available from the author on request.

The estimated cost functions can be used to predict and compare the costs of private and public production for each collection area. In other words, I compute $\hat{C}_{pr} - \hat{C}_{pu} = (\hat{\beta}_{pr} - \hat{\beta}_{pu})$. This suggests that private production on average is 6 per cent more costly than public production. The estimated mean of excess private costs, 0.064, has a standard error of 0.024 and is, therefore, significantly different from zero. This result is consistent with the finding, reported in Table 4.5, that municipalities that have used yardstick competition, in the sense that they have done cost analyses, are more likely to choose public production.

4.5.3.1 Pooling

I have calculated a likelihood ratio test of the hypothesis that the output coefficients are the same for private and public firms. Table 4.5, column 3, reports a dummy-variable estimation for the whole sample, controlling for sample selection, with the restrictions on output coefficients and on the sample selection-term coefficient imposed. The likelihood ratio test of the restrictions gives χ^2 (5) statistic with a value of 41.2, which corresponds to a significance level of 0.000. This test strongly rejects the pooling restrictions.

Which variables cause the pooling restriction to fail? The distance variable is significantly different in the private and public cost functions. Pick-up points per tonne is also borderline significantly different. In addition, the estimated coefficients for the selection terms imply different selection processes.[23]

There are examples in the previous refuse collection literature where separate cost functions are estimated and pooling tested: Gradus and Dijkgraaf (1997) reject pooling, while Stevens (1978) cannot reject pooling. Pier, Vernon, and Wicks (1974) estimate separate cost functions.

4.5.3.2 Logical Consistency

A consequence of the pooling tests is that the dummy-variable specification is rejected. This implies that there is no problem of logical inconsistency for these data.

Even if the dummy-variable model had passed the pooling tests, these data would still not give rise to a logically inconsistent model. The reason is that it turns out that cost differences do not matter for choice.

4.5.4 Producer Choice Taking Costs into Account

Table 4.6 suggests that cost differences do not matter for choice. Column 1 reports the estimation of a reduced form probit. The only significant output variables are pick-up points per tonne and quantity. I have computed excess private costs as previously described. Cost-minimizing behavior would imply that excess private costs

[23] The estimations are available from the author on request.

Table 4.6 Producer choice taking costs into account and test for cost minimization probit models

	Reduced form (1)	Structural form (2)
Socialist	0.09	0.22
	(0.28)	(0.77)
Share of single-family houses	1.96	2.87
	(1.20)	(1.95)
Competitive tendering, sealed bids	0.22	0.42
	(0.63)	(1.33)
Cost analysis made by the procuring municipality	-0.48	-0.64
	(1.64)	(2.46)
Average income	3.15	0.09
	(1.07)	(0.04)
Population	0.68	-0.48
	(0.74)	(0.70)
Population density	9.20	2.67
	(0.97)	(0.38)
Housing density	-16.80	-2.11
	(0.94)	(0.16)
Excess private costs		0.05
		(0.10)
Quantity	-1.07	
	(4.20)	
Pick-up points per tonne	-0.58	
	(2.00)	
Pick-up frequency	-1.12	
	(1.54)	
Distance	0.16	
	(0.98)	
Constant	2.13	-1.29
	(0.58)	(0.49)
Log likelihood	-52.95	-70.79
Average likelihood	0.65	0.56
χ^2	56.74	21.06
Significance level	0.000	0.012
Number of observations	122	122

Note: Absolute t-values are given in parentheses.

had a negative coefficient in the choice equation. Column 2 in Table 4.6, however, reports an insignificant (and positive) coefficient.

The result that municipalities that have analyzed costs are more likely to choose public production still holds in the structural probit. The share of single-family houses continues to have a positive impact on the probability of private production in the structural probit.

I have also tried to estimate simultaneously the cost models and the choice equation assuming that lower costs matter for choice. However, the maximum likelihood estimations either did not converge or gave unreasonable results. I interpret this as meaning that this model specification is not appropriate for the present data.

4.5.5 Model Selection

Table 4.7 summarizes the different results reported on the effect of ownership. Model selection is crucial. Suppose that we erroneously pool the data and do not correct for selectivity by using the dummy-variable model.

This gives the opposite result to a correctly specified model: costs in public firms are estimated to be 13 per cent higher than those in private firms. If we instead correct for selectivity but still erroneously pool the data, the public cost advantage is overestimated: costs in public firms are estimated to be 36 per cent lower than those in private firms. Pooling is less of a problem than not correcting for selectivity. Suppose that we correctly do not pool but erroneously do not correct for selectivity. We will get a result (14 per cent excessive public costs) opposite to the true result (6 per cent excessive private costs).

Table 4.7 Excess private costs (per cent)

	Separate cost models no pooling	Dummy-variable model pooling
Correction for selectivity bias	6.4[a]	35.9[b]
No correction for selectivity bias	−13.9[c]	−13.3[d]

[a] Computed using the estimations reported in Table 4.5, columns 1 and 2.
[b] Table 4.5, column 3, reports the estimation.
[c] Estimations available from the author on request.
[d] Table 4.3, column 1, reports the estimation.

4.6 Concluding Remarks

There are many studies of the possible differences in costs between private and public firms. However, few studies focus on what determines the political decision-makers' choice between private and public firms. Understanding this is as important as understanding possible cost differences.

Many comparisons of the performance of public and private firms use a public/private ownership dummy variable to capture cost differences in cross-section data. When estimating cost functions, the dummy variables are intended to capture the cost difference between, for example, public and private firms. The model has, however, nothing to say about producer choice. In addition, this approach raises three important empirical issues: pooling, selectivity and logical consistency.

If public and private firms use different production technologies, the dummy-variable approach is misspecified. Secondly, selectivity bias may arise, making it more difficult to identify cost differentials that actually exist. Thirdly, if data should be pooled, the resulting empirical model may be logically inconsistent.

I have compared public and private firms using refuse collection costs in 115 Swedish municipalities. In some municipalities, several firms collect in different areas. The data cover 170 firms.

In our case, it turns out that it is crucial to adjust for selectivity and that it is not correct to pool data for private and public firms. The dummy-variable model is misspecified. As data should not be pooled in this case, the problem of logical consistency does not arise. The two main results are as follows:

- The estimations suggest that public production, on average, was 6 per cent cheaper than private production.
- Cost differences did not affect producer choice. The municipalities, in other words, did not choose the least-cost alternative.

It is a main finding that policy-makers do not minimize costs. This is consistent with what I found in Ohlsson (1996), public firms pay more for their trucks. It is also consistent with the idea that private firms are likely to have higher capital costs than public firms. Private firms are also more likely to pay more for inputs. To select private firms, therefore, does not give the lowest costs. It is, finally, consistent with the argument that privatization, procurement and contracting out force public sector managers and procurers to adopt commercial criteria.[24]

The important conclusion is, therefore, that the main problem in the provision of refuse collection was that public policy-makers did not minimize costs, not that public firms were less efficient than private. They were not.

Acknowledgments The Swedish Competition Authority financed the research and also collected and provided most of the data. The author thanks Sören Blomquist, Pierre Pestieau and two anonymous referees for helpful comments and suggestions. He has also benefited from comments by seminar participants at Uppsala University, Göteborg University, Umeå University and the Norwegian University of Science and Technology, Trondheim. Some of the work on this chapter was carried out while the author enjoyed the hospitality of ERMES, Université Panthéon-Assas, Paris II. This chapter was published as Ohlsson H. (2003) Ownership and production costs: Choosing between public production and contracting out in the case of Swedish refuse collection. Fiscal Studies 24:451–476.

Appendix 1: The Data

Municipality-Level Data

Socialist = 1 if the Social Democrats and the Left Party together had a majority of the seats in the municipality council, 1987. Source: Statistics Sweden, *Yearbook for Swedish Municipalities 1987.*

Share of single-family houses: The number of single-family houses divided by the total number of housing units, 1985. Source: Statistics Sweden, *The 1985 Census.*

Average income: Total factor income per inhabitant older than 20 years, 1987. Source: Statistics Sweden, *Yearbook for Swedish Municipalities 1989.*

---

[24] This argument is made by, among others, Haskel and Szymanski (1993).

Population, 1 January 1987. Source: Statistics Sweden, *Yearbook for Swedish Municipalities 1987.*

Population density: The number of inhabitants per square kilometer, 1 January 1987. Source: Statistics Sweden, *Yearbook for Swedish Municipalities 1987.*

Housing density: Total number of housing units in the municipality, 1985, divided by the area of the municipality. Sources: Statistics Sweden, *The 1985 Census* (housing units); Statistics Sweden, *Yearbook for Swedish Municipalities 1987* (area).

Firm-Level Data

The firm-level data come from the 1989 survey by the Swedish Competition Authority, formerly the Swedish National Price and Cartel Office, concerning 1987.

Competitive tendering, sealed bids: When collection in the area is currently done by a public firm, data are from Survey Form C, question 8. When the municipality currently contracts out collection in the area to a private firm, data are from Survey Form C, question 11, column 6.

Cost analysis made by the procuring municipality: When collection in the area is currently done by a public firm, data are from Survey Form C, question 9. When the municipality currently contracts out collection in the area to a private firm, data are from Survey Form C, question 11, column 9.

Quantity: Data are in tonnes and come from the answers to Survey Form E, question 2.

Pick-up points per tonne: Data on the number of pick-up points come from the answers to Survey Form B, question 1. I have then divided the data by quantity.

Pick-up frequency: Data on number of pick-ups come from the answers to Survey Form B, question 1. I have then divided the data by the number of pick-up points, data on which come from the answers to Survey Form B, question 1.

Distance: Data for distance driven are in kilometers and come from Survey Form B, question 7.

Costs: The sum of the costs for labor, vehicles, offices, packing and some minor cost items. Data are in SEK per tonne and come from Survey Form D, lines 6–21 costs + lines 22–24 depreciation – line 19 payments to contractors. Sometimes, both the municipality and the firm have costs for collection within a certain area. In these cases, I have added the costs, but payments to contractors should not be included.

Wage rate: Wage costs divided by the number of employees. Wage cost data including payroll taxes are in SEK and come from Survey Form D, lines 6 and 7. Data on the number of employees (full-time all-year equivalents) are from Survey Form B, question 4.

Cost of capital: Vehicle costs divided by the number of vehicles. Vehicle cost data are in SEK and come from Survey Form D, line 10 leasing costs, line

11 repairs and fuel, line 13 insurance, line 14 taxes and line 22 depreciation. Data on the number of vehicles are from Survey Form B, question 7.

Private ownership: Data are from the list of identification codes of firms and municipalities.

Appendix 2: Maximum Likelihood Estimations

Table 4.8 Cost functions: sample selection model

	Private firms		Public firms		All firms	
	Coef.	z-value	Coef.	z-value	Coef.	z-value
Cost function						
Quantity	−0.14	1.80	−0.07	0.96	−0.09	1.67
Pick-up points per tonne	0.16	1.80	0.46	4.63	0.21	3.11
Pick-up frequency	−0.01	0.07	0.24	1.05	0.01	0.14
Distance	0.22	3.14	0.05	1.41	0.16	3.51
Private ownership					0.41	2.48
Constant	−1.47	2.22	−1.18	1.81	−1.74	3.20
Producer choice						
Socialist	0.07	0.27	−0.31	0.64	−0.04	0.18
Share of single-family houses	2.48	1.78	3.35	1.52	1.52	1.20
Competitive tendering, sealed bids	0.31	1.03	−0.42	1.26	0.21	0.82
Cost analysis made by procuring municipality	−0.50	1.97	0.64	2.15	−0.38	1.74
Average income	0.73	0.32	0.12	0.04	0.63	0.32
Population	−1.03	1.46	0.15	0.11	−1.49	2.29
Population density	−0.82	0.13	−4.20	0.45	−2.94	0.52
Housing density	5.67	0.46	5.80	0.29	10.80	0.98
Constant	−1.47	0.61	1.51	0.57	−0.63	0.29
$\sigma_{u_i v}$	−0.60	2.89	−0.40	0.25	−0.71	5.88
σ_{u_i}	0.61	8.95	0.24	2.44	0.54	10.7
Log likelihood	−131.65		−67.40		−147.64	
χ^2	15.77		30.10		33.13	
Significance level	0.003		0.000		0.000	
No. of uncensored observations	75		47		122	
No. of observations	122		122		122	

Note: All output variables are in logarithms. Z-values in absolute terms.

References List

Bennett, J. T., & Johnson, M. H. (1979). Public versus private provision of collective goods and services: garbage collection revisited. *Public Choice*, *34*, 55–63.

Bennett, J. T., & Johnson, M. H. (1980). Tax reduction without sacrifice: private-sector production of public services. *Public Finance Quarterly*, *8*, 363–96.

Bivand, R., & Szymanski, S. (2000). Modelling the spatial impact of the introduction of compulsory competitive tendering. *Regional Science and Urban Economics*, *30*, 203–19.

Borcherding, T. E. (1988). Some revisionist thoughts on the theory of public bureaucracy. *European Journal of Political Economy, 4*, 47–64.
Borcherding, T. E., Pommerehne, W. W., & Schneider, F. (1982). Comparing the efficiency of private and public production: the evidence from five countries. *Zeitschrift für Nationalökonomie Supplement 2*, 127–156.
Bös, D. (1991). *Privatization: A theoretical treatment.* Oxford: Oxford University Press.
Bosch, N., Pedraja, F., & Suàrez-Pandiello, J. (2000). Measuring the efficiency of Spanish municipal refuse collection services. *Local Government Studies, 26*, 71–90.
Collins, J. N., & Downes, B.T. (1977). The effects of size on the provision of public services: the case of solid waste collection in smaller cities. *Urban Affairs Quarterly, 12*, 333–347.
Cubbin, J., Domberger, S., & Meadowcroft, S. (1987). Competitive tendering and refuse collection: Identifying the sources of efficiency gains. *Fiscal Studies, 8*, 49–58.
Davidson, R., & MacKinnon, J. G. (1993). *Estimation and inference in econometrics.* Oxford: Oxford University Press.
De Silanes, F. L., Schleifer, A., & Vishny, R. W. (1997). Privatization in the United States. *RAND Journal of Economics, 28*, 447–471.
Dijkgraaf, E., Gradus, R. H. J. M., & Melenberg, B. (2003). Contracting out of refuse collection. *Empirical Economics, 28*, 553–570 (reproduced in chapter 3).
Domberger, S., & Jensen, P. (1997). Contracting out by the public sector: theory, evidence, prospects. *Oxford Review of Economic Policy, 13*, 67–78.
Domberger, S., Meadowcroft, S. A., & Thompson, D. (1986). Competitive tendering and efficiency: The case of refuse collection. *Fiscal Studies, 7*, 69–87.
Dubin, J. A., & Navarro, P. (1988). How markets for impure public goods organize: The case of household refuse collection. *Journal of Law, Economics and Organization, 4*, 217–241.
Edwards, F. R., & Stevens, B. J. (1978). The provision of municipal sanitation services by private firms: an empirical analysis of the efficiency of alternative market structures and regulatory arrangements. *Journal of Industrial Economics, 27*, 133–147.
Ferris, J. M. (1986). The decision to contract out: an empirical analysis. *Urban Affairs Quarterly, 22*, 289–311.
Gómez-Lobo, A., & Szymanski, S. (2001). A law of large numbers: Bidding and compulsory competitive tendering for refuse collection contracts. *Review of Industrial Organization, 18*, 105–113.
Gradus, R. H. J. M., & Dijkgraaf, E. (1997). Costs savings of contracting out refuse collection., Research Memorandum 9712, Erasmus University Rotterdam, Rotterdam (published as Dijkgraaf E, Gradus RHJM (2003). Cost savings of contracting out refuse collection. *Empirica 30*, 149–161 and reproduced in chapter 2).
Greene, W. H. (1995). *LIMDEP*. Bellport, New York: Econometric Software Inc.
Haskel, J., & Szymanski, S. (1993). Privatisation, liberalisation, wages and employment: Theory and evidence. *Economica, 60*, 161–181.
Heckman, J. J. (1978). Dummy endogenous variables in a simultaneous equation system. *Econometrica, 46*, 931–959.
Hirsch, W. Z. (1965). Cost functions of an urban government service: refuse collection. *Review of Economics and Statistics, 47*, 87–92.
Hoover, G. A., & Peoples, J. (2003). Privatization of refuse removal and labor costs. *Journal of Labor Research, 24*, 293–305.
Kemper, P., & Quigley, J. M. (1976). *The economics of refuse collection.* Cambridge, MA: Ballinger Publishing Company.
Kitchen, H. M. (1976). A statistical estimation of an operating cost function for municipal refuse collection. *Public Finance Quarterly, 4*, 56–76.
Maddala, G. (1983). *Limited dependent and qualitative variables in econometrics.* Cambridge, MA: Cambridge University Press.
Maddala, G., & Lee, L. (1976). Recursive models with qualitative endogenous variables, *Annals of Economic and Social Measurement, 5*, 525–545.

Marcinèin, A., & van Wijnbergen, S. (1997). The impact of Czech privatization methods on enterprise performance incorporating initial selection-bias correction. *Economics of Transition, 5*, 289–304.
McDavid, J. C. (1985). The Canadian experience with privatizing residential solid waste collection services. *Public Administration Review, 45*, 602–608.
Nelson, M. A. (1997). Municipal government approaches to service delivery: an analysis from a transactions cost perspective. *Economic Inquiry, 35*, 82–96.
Ohlsson, H. (1996). Ownership and input prices: a comparison of public and private enterprises. *Economics Letters, 53*, 33–38.
Pestieau, P., Tulkens, H. (1993). Assessing and explaining the performance of public enterprises. *FinanzArchiv, 50*, 293–323.
Pier, W. J., Vernon, R. B., & Wicks, J. H. (1974). An empirical comparison of government and private production efficiency. *National Tax Journal, 27*, 653–656.
Pommerehne, W. W., & Frey, B. S. (1977). Public versus private production efficiency in Switzerland: a theoretical and empirical comparison. In V. Ostrom & F. P. Bish (Eds.), *Comparing urban service delivery systems, Urban Affairs Annual Review* (Vol. 12). California, CA: Sage Beverly Hills.
Reeves, E., & Barrow, M. (2000). The impact of contracting out on the costs of refuse collection services: The case of Ireland. *Economic and Social Review, 31*, 129–150.
Savas, E. (1977a). An empirical study of competition in municipal service delivery. *Public Administration Review, 37*, 717–724.
Savas, E. (1977b). Policy analysis for local government: public vs. private refuse collection. *Policy Analysis, 3*, 49–74.
Savas, E. (1981). Intracity competition between public and private service delivery. *Public Administration Review, 41*, 46–52.
Sonenblum, S., Kirlin, J. J., & Ries, J. C. (1977). *How cities provide services: An evaluation of alternative delivery strategies*. Cambridge, MA: Ballinger Publishing Company.
SPK (Swedish National Price and Cartel Office) (1991). Sophämtning i kommunal och privat regi. Marknadsstruktur, konkurrens och kostnader. SPKs Rapportserie no. 1991:6.
StataCorp (2001). Stata Statistical Software: Release 7.0. Stata Corporation, College Station, TX.
Stevens, B. J. (1978). Scale, market structure, and the cost of refuse collection. *Review of Economics and Statistics, 60*, 438–448.
Szymanski, S. (1996). The impact of compulsory competitive tendering on refuse collection services. *Fiscal Studies, 17*, 1–19.
Szymanski, S., & Wilkins, S. (1993). Cheap rubbish? Competitive tendering and contracting out in refuse collection, 1981–1988. *Fiscal Studies, 14*, 109–130.
Tang, K. (1997). Efficiency of the private sector: a critical review from empirical evidence from public services. *International Review of Administrative Sciences, 63*, 459–474.
Tickner, G., & McDavid, J. C. (1986). Effects of scale and market structure on the costs of residential solid waste collection in Canadian cities. *Public Finance Quarterly, 14*, 371–393.
Vickers, J., & Yarrow, G. (1988). *Privatization: An economic analysis*. Cambridge, MA: MIT Press.
Vining, A. R., & Boardman, A. E. (1992). Ownership versus competition: efficiency in public enterprise. *Public Choice, 73*, 205–239.
Wittman, D. (1989). Why democracies produce efficient results. *Journal of Political Economy, 97*, 1395–1424.

Chapter 5
Does Public Ownership Impair Efficiency in Norwegian Refuse Collection?

R.J. Sørensen

Abstract Corporate governance theory suggests that companies with dispersed and indirect ownership suffer from agency costs. A worst case is where several political authorities jointly own a company, which allows managers to operate with inferior efficiency. In political economy, the manager is not the major agency problem. Elected politicians may impair efficiency to improve their re-election prospects. Since politicians have less influence in jointly owned firms, such companies are expected to perform better than those owned by a single public authority. Consistent with corporate governance, but not political economy, the empirical analysis suggests that dispersed municipal ownership impairs cost efficiency. In the Norwegian case of municipal refuse collection presented here, costs of dispersed ownership often outstrip gains from economies of scale. Use of jointly owned companies is not necessarily a proper response to efficiency problems inherent a fragmented local government structure.

Keywords Dispersed ownership · efficiency · agency costs · collection · Norway

5.1 Introduction

Comparing public and private organizations has become a big industry. Researchers have devoted less attention to the efficiency of different types of public service organization. Consider the intermunicipal company, which, in many countries, has become an important organizational entity. First, small local governments are often unable to exploit economies of scale. In many cases, two or more neighboring municipalities set up a jointly owned corporation, an intermunicipal company. Such organizations can take advantage of economies of scale in infrastructure sectors such as refuse collection and disposal, water supply and sewage treatment, and electricity

✉ R.J. Sørensen
Department of Public Governance, Norwegian School of Management, Nydalsveien 37, 0484 Oslo, Norway
e-mail: rune.sorensen@bi.no

E. Dijkgraaf, R.H.J.M. Gradus (eds.), *The Waste Market*,

distribution. Second, local authorities are increasingly applying competitive tendering or other forms of market competition to provide infrastructure services. Intermunicipal companies are subject to the same legal framework as private companies. They are also better equipped to compete than conventional public agencies. Management and board have considerably greater discretion than leaders of government agencies.

Corporate governance theory suggests that such dispersed ownership creates a collective action problem, which can lead to a loss of ownership control and inferior performance. Public ownership is in itself an extreme case of ownership dispersion. All citizens have a share in the company! Since individual voters lack the power to monitor their agents directly, ownership rights must be exercised through elected representatives. Corporate governance suggests that indirect ownership yields lower efficiency than direct ownership. Ownership control becomes further diluted when more than one political authority controls a company. Intermunicipal companies may therefore have lower cost efficiency than companies owned by a single authority. In principle, the efficiency gains related to scale economics could be smaller than the efficiency loss due to multiple owners.

Political economy offers a completely different story. Where corporate governance theorists consider active owner-representatives to be an asset, political economists see the active politician as the problem. Suppose owner-citizens lack the information required to oversee the management in publicly owned companies. In these companies, voters do not necessarily punish elected representatives for lack of efficiency. Local politicians can therefore use the municipal company to cater for company employees or other important groups of voters. If management in companies owned by several municipalities is more shielded from political pressure than administration companies owned by a single municipality, we should expect the jointly owned corporation to have better cost efficiency than firms owned by a single local government.

This chapter explores these propositions by means of data on Norwegian local government. Looking at the issue of refuse collection and disposal, the empirical analysis suggests that it is the number of government owners that exerts a negative influence on unit costs. The efficiency losses induced by fragmented ownership often exceed the gains of operating on a larger scale. In the section that follows, we elaborate on theoretical perspectives. We then describe the institutional setting, outline the research design and present empirical results on the role of ownership on costs and user fees for refuse collection.

5.2 Corporate Governance Versus Political Economy

Since the early 1990s, there has been a marked interest in issues related to corporate governance in both public and private sector settings. Its basic premise is that a runaway management weakens company performance, and that active owners are desirable to sustain efficiency and profitability. On the other hand, political economy suggests that active politician-owners are the essence of the problem, not the solution. We provide a brief review of these theories and suggest a way to discriminate empirically between the two conflicting propositions.

5.2.1 Corporate Governance Theory

Agency theory forms the benchmark model of corporate governance. Delegation of ownership rights may improve performance if agents are more competent than principals, but delegation may also entail a loss of control. Concentrated ownership strengthens incentives to oversee company management, which is expected to yield a positive net effect on performance.

First, dispersed ownership means that each owner has a weak incentive to monitor the performance of company leaders. Lack of collective action among principals leads ownership to become separated from control (Fama & Jensen, 1983, 309). Second, owners will search for institutional alternatives that compensate for lack of monitoring of company management. One such mechanism is economic incentives. However, multiple owners do not necessarily have identical interests, something that creates a common agency problem. Since incentives to reach one goal may undermine other goals, the overall result may easily be diluted incentives. Dixit (1997) suggests that public agencies must answer to more constituencies than do privately owned organizations. Finally, corporate governance theory observes that government ownership represents a polar case of dispersion. Even in relatively small local governments, ownership controls must be delegated to administrators. All citizens have ownership rights, but very few have control rights. Since administrators pursue objectives that differ from the goals of the owner-citizen, publicly owned companies are expected to perform relatively poorly (Shleifer & Vishny, 1997).

According to corporate governance theory, the worst case is a combination of indirect and dispersed ownership. One such example is the case where numerous institutional investors (pension funds, insurance companies, and so on) own an entire private company. Another example is the case considered here: multiple political authorities (municipalities) exercise ownership rights on behalf of their citizens. Corporate governance theorists suggest that intermunicipal companies should have the weakest performance of all institutional creations.

5.2.2 Political Economy Theory

In principle, in their management of publicly owned companies, we would not expect elected politicians to behave differently from private owners. Inferior performance would imply higher costs than necessary, which would lead to either higher taxes or lesser revenue available for other political purposes. In other words, a politician should seek to minimize costs for a given service output and a given quality in order to maximize electoral support. When citizens are unable to oversee their elected representatives, politicians can get additional voter support by deviating from efficient operation of the company. Active political ownership may therefore undermine profitability and cost efficiency in publicly owned companies (Shleifer & Vishny, 1994).

In the case of a publicly owned firm, excess employment is one potential source of inefficiency. Local politicians have an incentive to intervene in the operation of such a company for the Benet of its employees (Boycko, Shleifer, & Vishny, 1996), since they are more likely to support incumbent parties that protect the firm. On the other hand, inefficiency diminishes the profitability of the publicly owned firm, which reduces the welfare of other groups of voters. These voters are unlikely to be either informed about or concerned about their loss of profitability in a government firm. In the case of a privately owned firm, in a competitive environment, explicit subsidies must be used to maintain excess employment. Voters not employed by the private firm are likely to become informed and alarmed about such cash transfers. This reduces or eliminates the political gain of surplus employment in a privately owned firm.

In the case addressed in the current empirical analysis, a public authority purchases services from a publicly owned firm. This means that 'invisible ports' and 'visible subsidies' are less relevant. In either case, inefficient service provision leads to either higher taxes or fees or fewer resources are available for providing other public services. Suppose one municipality is the sole owner of a company. When information problems hinder electoral controls, elected politicians can be tempted to purchase from the firm even if it is less than efficient. Incumbent politicians can gain votes from company employees without losing voter support from other citizens.

Compare this situation with one where the company is owned by two or more municipalities. Suppose that facilities have been located in one of the municipalities to minimize costs. Employees living and working in this municipality are likely to resist demanding efficiency initiatives, particularly if such programmes involve personnel reductions (Shleifer & Vishny, 1994). Note that infrastructure services are not labor intensive, and that the surplus employment renders cost per taxpayer and per voter quite insignificant and 'invisible'. Local politicians may support the opposition of local employees to further their re-election prospects. Elected politicians from other municipalities are likely to dispute this, and support efficiency improving programmes. When decisions are made with majority voting in the company's general assembly, proposals that will harm efficiency are not likely to get majority support. This presupposes that a dominant owner is incapable of exploiting the others for its particular political purposes. In other words, governmental intervention entails higher transaction costs under intermunicipal ownership than under a single municipal owner (Sappington & Stiglitz, 1987). In direct contrast to corporate governance theory, the intermunicipal company should be more efficient than companies owned by a single municipality.

The two hypotheses, corporate governance and political economy, are not necessarily incompatible. Both governance problems can arise at the same time. Elected politicians may not speak for the interests of a majority of citizen-principals, while at the same time corporate managers are imperfect agents of their politician-principals.

We can identify which governance problem is more serious, administrative autonomy or politician control, by examining how number of local government owners affects company performance.

5.2.3 Empirical Studies on Public Ownership

Ownership dispersion has been studied extensively in the corporate governance literature. Becht, Bolton and Roell (2003, 63) have identified several generations of empirical research that addresses this hypothesis. Most show that portability is higher in owner-controlled companies and family owned firms (see for example Bøhren & Ødegaard, 2005). Other empirical studies use indicators of ownership dispersion. The empirical results are not conclusive: several studies suggest that dispersion leads to inferior performance, while a lot of other papers reject the hypothesis.

Empirical studies related to boards of directors have often produced inconclusive results. The message appears to be that boards in many cases are inefficient substitutes for active and concentrated ownership. They are commonly seen as inefficient supervisors of the CEO, sometimes even 'captured' by company management. Taking other governance mechanisms into account, hostile takeovers, large shareholders and CEO incentives, and so on, corporate governance research cannot be said to have shown particularly robust results (Becht et al. 2003, 83)

A relatively large literature addresses the impact of public and private ownership in the contexts of both monopoly and competition in the management of waste. For example, Savas (2000) asserts that the best way of organizing garbage collection is to divide the jurisdiction into appropriate sections and organize competitive bidding for the sections from private firms and municipal agencies. The meta-study of Domberger and Jensen (1997) suggests that most frequently reported cost reductions from contracting out are between 10 and 30 per cent. This appears not to result from reductions in wage levels, but from a broad set of managerial initiatives to improve cost performance. Waste collection has probably been studied more extensively than any other service. For example, Gomez-Lobo and Szymanski (2001) investigate UK local authority refuse collection contracts, and find that a higher number of bids are associated with significantly lower cost of service. In The Netherlands, Dijkgraaf & Gradus (2003) find that contracting out yields cost savings of about 15–20 per cent.

Competition often, but not always, reduces efficiency differences between public and private firms. In the Danish dental sector, Andersen and Blegvad (2006) observe no differences in cost-efficiency or effectiveness between public and private producers. Hjalmarsson and Veiderpass (2002) analyze regulated local monopolists providing electricity distribution in Sweden. This study reveals no significant differences in efficiency and productivity growth between private and public companies. Caves and Christensen (1980) compared two large public and private railroad companies in Canada, which competed over many routes. Initially, the private company had higher productivity than the public company. These differences were soon eliminated, and productivity differences disappeared. Borcherding, Pommerehne, and Schneider (1982), Domberger and Jensen (1987) and D'Souza and Megginson (1999) provide further evidence and extensive reviews on the impact of public and private ownership.

Dubin and Navarro (1988) provide a rare example of empirical research on the political economy of government ownership. They analyze alternative governance systems of refuse collection in the US setting. They argue that a proper specification of the cost functions, particularly the role of density, is important for assessing the role of alternative systems of garbage collection. Their analysis suggests that private market organization (that is, no government regulation of refuse collection) is significantly more costly than contracting, franchise or municipal provision (see also Vining & Boardman, 1992). This system fails to take into account economies of density in garbage collection. Both municipal provision and contracting is more efficient than private market organization, while franchise is in between (Dubin & Navarro, 1988, 233).

What is striking is the almost total lack of research addressing performance differences between organizations operating within the public sector (Dunsire, Hartley, Parker & Dimitriou, 1988, 368). One exception is the observation that corporatization of public enterprises has a modest disciplining impact on ownership governance and organizational performance (Dunsire, Hartley & Parker, 1991; Shirley 1999). In light of the conflicting theoretical propositions outlined above, it is particularly remarkable that no empirical study has analyzed the performance of companies with more or less dispersed 'public' ownership.

5.3 Refuse Collection in Norwegian Local Government

Governments increasingly provide public services by means of publicly owned companies. Throughout Europe, national governments own companies which provide postal services, telecommunications, electricity, and public transportation. In local government, the number of such companies has increased considerably in the infrastructure sectors. In the Norwegian context, municipal companies distribute electricity in local and regional networks, provide parking facilities, manage municipal properties, operate ports, provide water and sewerage services, and collect, handle and dispose of household and business waste (Sørensen & Bay, 2002). The number of companies owned by Norwegian local authorities has increased from 1560 companies in 1999 to 2203 in 2004. Numbers of firms that are independent legal entities have increased from 773 companies in 1996 to 1728 in 2004 (Statistics Norway, 2004).

Garbage collection and disposal is one of the most intensely researched infrastructure services. As a relatively simple public service, it is frequently considered well suited for competitive tendering and outsourcing. Furthermore, the European Union has implemented a number of regulations designed to impose more competition into the waste management market. Despite emphasis on competitive regulation and competitive tendering, private contractors' market share varies considerably across countries. For example, 80 per cent of garbage collection in Spain has been outsourced to private contractors, 60 per cent in Germany and 50 per cent in France. However, this applies to only 30–40 per cent in the UK and The Netherlands and 10–15 per cent in Norway (Hall, 2006). In the Norwegian case, the number refers to percentage of municipalities that purchase refuse services from private companies. Thus, governments continue to play an important role in the organization of waste collection and treatment in many countries.

Privately owned companies in the waste industry often expand by mergers and acquisitions. Government organizations appear to follow in their footsteps. However, such consolidations in local government often face intense popular opposition in at least one of the affected constituencies. Various forms of intermunicipal alliances are politically attractive, since it avoids the political costs of dismantling existing political institutions. Similar to other countries (Dunsire et al., 1988, 366–367), local governments in Norway have established companies to take advantage of scale economies. In some cases, they set up traditional limited liability companies, which are owned by one or more municipalities, possibly with private owners as well. The entity considered here is called the intermunicipal corporation. It has unlimited liability, but can only be owned by two or more local governments. Number of intermunicipal companies has increased from 7 in 1996 to 206 in 2004.

The empirical analysis presented here employs data about refuse collection and treatment, which in the Norwegian context is a municipal responsibility. Local government in Norway comprises 434 municipalities and 18 counties. It should be noted that data refer to the year 2005 and do not refer to Oslo. Local elections to municipal and county councils are held every four years in between national elections. Municipalities have responsibility for establishing and operating many things: kindergartens, primary schools, health centers/primary health services, social welfare, culture (cinema, sports, music schools, and so on), some clerical functions, communication (municipal roads), infrastructure services (including water works, sewers, refuse collection and disposal), planning and construction, industry development, public utilities and tax collection.

Tax revenues account for 45 per cent of municipal revenues. Most of the tax revenues are collected as a proportional payroll tax, that is, as an income tax. Central government stipulates the minimum and maximum levels of tax rates. In Norway, all municipalities use the maximum tax rates. Property taxes play a minor role. Block grants and earmarked grants from central government account for most of the other revenues. Exogenous per capita revenues include block grants from central government plus revenues from income and asset taxation.

Municipalities collect user fees as well. In the refuse sector, fees are legally required to cover the costs of providing the service (PPP: 'Polluter pays principle'). Nevertheless, local governments have considerable discretion in how they stipulate the unit costs of collecting and handling refuse. Local governments may choose either to subsidize refuse collection to reduce user fees or they may use the fees to finance other government services such as education or health care.

5.4 The Impact of Dispersed Ownership on Costs and User Charges

Garbage collection is a relatively simple production activity. Households and firms leave their garbage at collection points and the service operators transport the garbage to disposal sites. A service operator basically needs drivers, loaders and collection vehicles. The potential cost of garbage collection depends on several factors such as regional characteristics (for example, density of collections points), service specifications (for example, sorting of garbage, frequency of collection), productivity of labor and capital, and input prices.

In Table 5.1, we present relevant descriptive statistics for the multivariate analyses. The table comprises data on two performance variables collected by Statistics Norway: (1) yearly fees for refuse collection and handling for a standard household; and (2) total direct and indirect costs derived of refuse collection and handling, measured per capita. Both user fees per household and total costs per inhabitant decrease with centrality. As is to be expected, refuse collection can be provided at lower costs and lower prices in central areas (see Box 5.1).

We utilize three measures of ownership concentration/dispersion: (1) Herfindahl index of ownership concentration, which is commonly applied in the corporate governance literature; (2) number of municipal owners; and (3) whether or not the municipality cooperates with other authorities in the refuse sector. Data on ownership structures has been taken from the official Register of Legal Entities (the Brønnøysund Register), and data on other types of intermunicipal cooperation has been derived from a government database on local government organization. Average number of municipal owners of these intermunicipal companies are 6.8 owners, which appears quite high. The average municipality (including those that operate the service alone) provides refuse services through an organization owned by seven other local governments. The Herfindahl index of ownership concentration is low, with an average of 0.35.

Table 5.1 also contains information on four variables used as controls in the subsequent regression analyses. First, municipal revenues comprise block grants and taxes on income and assets. Due to the regulation of these tax rates and the fact that all municipalities have for several decades used the maximum rates, municipal revenues per capita can be considered exogenous in this context. Local governments that are centrally located have lower per capita tax revenues (income and asset taxes) than peripheral authorities. Central government allocates much higher block grants to rural municipalities. Total local government revenues (other than user fees and small revenues from property taxes) are therefore significantly higher in peripheral municipalities than in central areas.

Second, since transportation costs are an important component of refuse collection, we include information about shares of population living in sparsely and densely populated areas. Densely populated areas have at least 200 inhabitants per settlement, and the distances between houses are no more than 50 meters. As to be expected, settlement patterns are denser in centrally located municipalities.

Table 5.1 Descriptive statistics on refuse collection and handling in Norwegian municipalities, 2005 (averages by municipal centrality)

Municipal centrality	Fees per household (NOK)	Costs per capita (NOK)	Owners	Herfindahl	Cooperation	Competition	Revenue per Capita (NOK)	Share of population in sparsely populated areas, 2005	Cooperating population	Municipal population	N*
1.	1992	1010	7.5	0.25	0.93	0.29	34,500	0.60	32,290	2,895	164
2.	1948	896	6.3	0.37	0.80	0.41	30,290	0.65	40,881	3,881	35
3.	2154	850	7.8	0.29	0.88	0.21	28,887	0.40	35,706	6,807	25
4.	1704	742	4.9	0.41	0.81	0.19	27,676	0.49	31,934	7,927	21
5.	1901	845	8.1	0.30	0.92	0.20	27,027	0.40	53,620	9,660	37
6.	1701	780	4.3	0.51	0.61	0.50	25,826	0.39	59,471	17,965	44
7.	1764	675	6.3	0.44	0.86	0.39	25,092	0.35	112,047	23,753	104
All	1882	858	6.8	0.35	0.86	0.33	29,626	0.49	57,080	10,617	430
(Standard deviation)	(336)	(259)	(4.4)	(0.30)	(0.34)	(0.47)	(7042)	(0.27)	(65 202)	(30 639)	

Box 5.1 Municipal centrality (Standard classification of municipalities)

1. Less central municipalities and does not lie within 2.5 hours (in the case of Oslo: 3: hours) from an urban settlement on level 3(0B)
2. Less central municipalities and lies within 2.5 hours (in the case Oslo: 3: hours) from an urban settlement on level 3(0A)
3. Less remote municipalities and does not lie within 2.5 hours (in the case of Oslo: 3: hours) from an urban settlement on level 3 (1B)
4. Less remote municipalities and does lies within 2.5 hours (in the case of Oslo: 3: hours) from an urban settlement on level 3 (1A)
5. Fairly central municipalities and does no lie within 2.5 hours (in the case of Oslo: 3: hours) from an urban settlement on level 3 (2B)
6. Fairly central municipalities and lies within 2.5 hours (in the case of Oslo: 3: hours) from an urban settlement on level 3 (2A)
7. Central municipalities (3A)

Notes:
NOK: Norwegian Kroner
Fees: yearly refuse fee for a standard house of 120 square meters, exclusive value added tax. 1.1.2006
Costs: estimated costs (gross current costs + indirect cost + estimated costs on interest + misc. revenues) per capita, 2005
Owners: number of municipal owners of municipal or intermunicipal company or organization, 2005
Herfidahl: Herfindahl index of ownership concentration, 2005
Cooperation: coded 1 if the municipality cooperates with other municipalities in provided refuse collection and handling, and 0 otherwise
Competition: coded 1 if the municipality uses competitive tendering, and 0 otherwise
Revenues: local tax revenues (exogenous) and block grants, per capita, 2005
Cooperating population: population size covered by (inter) municipal cooperation/corporation, 2005
Municipal population: population size in municipality, 2005
N: number of municipalities

Third, market competition can induce optimal organizations. This interpretation is in line with the traditional conjecture that competition is more important than ownership (Bartel & Harrison, 1999). At least two observations suggest that this explanation is dubious in the current context. First, markets for refuse collection are far from perfect. At least in the Norwegian context, many local governments have not established a proper separation between the role of purchasing services and producing. A lack of regulatory transparency (that is, estimation of overhead costs) facilitates municipal cross-subsidization of in-house service provision. There are also examples of local governments that have awarded contracts to more expensive bidders, leading courts to rule against the municipality. Second, ownership effects can be identified in traditional markets (Vining & Boardman, 1992; Villalonga, 2000). As discussed above, number of owners affects efficiency and profitability for private companies operating in traditional competitive markets. In additional analyses (not presented), we included an interaction term between ownership dispersion and use of competitive tendering. We found no support for interaction, which implies that ownership dispersion and types do not impact differently under monopoly and competition. Table 5.1 shows that about 30 per cent of municipalities use competitive tendering to purchase services. These numbers suggest that ownership and management could have greater efficiency effects as compared to organizations that are subjected to more intensive market competition.

Fourth, we include population sizes to tap economies of scale. The relevant statistic includes the population size of region covered by the intermunicipal company to tap economies of scale or the municipal population when a single municipality produces services. For comparison purposes, Table 5.1 provides population number for each municipality. The least central municipalities have very small populations, with an average number of less than 3,000 inhabitants. The use of intermunicipal cooperation implies that these municipalities can use a single organization to provide refuse collection and handling to a population of more than 30,000 inhabitants. Intermunicipal cooperation is widespread in all types of municipalities, both central and peripheral authorities. Based on the data summarized in Table 5.1, we estimate regression models for user fees per household and costs per capita. We then assess the impact of ownership dispersion controlling for the factors displayed in Table 5.1. The regression estimates are shown in Table 5.2.

Ownership structures impact significantly on user fees and costs. When ownership is measured by the Herfindahl index (column I), estimates suggest that an increase in ownership concentration from 0 to 1 will reduce user fees by nearly 8 per cent and costs by 6 per cent. More municipal owners increase fees and costs (column II). An increase in number of owners from 1 to 6 (see Table 5.1) will increase user fees by about 10 per cent and costs by about 5 per cent. Finally, a dummy variable for use of intermunicipal cooperation (column III) yields similar results. Other factors being constant, local governments that cooperate with other authorities to provide refuse services have user fees and costs that are about 10 per cent higher than those that supply the service single-handedly.

It is interesting to compare these efficiency losses with the gains obtained by economies of scale. As can be seen in Table 5.2, the estimates diverge somewhat for user fees, but the cost-regressions suggest a scale elasticity of about –0,05. If intermunicipal cooperation yields an increase in the population base from 10,000 inhabitants to 60,000 inhabitants, the regression estimate suggest a reduction in costs of about 4 per cent. For many municipalities, governance losses due to dispersed ownership tend to exceed the gains from economies of scale.

Levels of municipal revenue have a significant positive effect on costs: a 1 per cent increase in per capita revenue increases costs by 0.4–0.5 per cent. Comparable results have been obtained for other public services in Norway,

Table 5.2 The impact of dispersed public ownership on unit costs and fees OLS regression

	I	II	III	I	II	II
Intercept	6.46***	6.82***	6.29***	5.70***	5.82***	6.60***
	(23.7)	(25.3)	(25.2)	(12.4)	(13.5)	(14.8)
Taxes and block grants per inhabitant (log)	0.355***	0.337***	0.399***	0.415***	0.442***	0.477***
	(5.11)	(5.19)	(6.05)	(3.73)	(4.37)	(4.77)
Competition (=1)	0.0012	–0.0047	–0.0049	0.014	–0.021	–0.26
	(0.06)	(0.25)	(–0.25)	(–0.38)	(–0.67)	(–0.79)
Share living in sparsely populated areas	–0.104*	–0.115*	–0.11	0.148*	–0.134*	–0.149*
	(–2.11)	(2.60)	(–2.42)	(–1.87)	(–1.91)	(–2.08)
Total population in cooperating municipalities (log)	0.0009	0.047**	–0.015	–0.040*	–0.066*	–0.058**
	(–0.66)	(–3.36)	(–1.24)	(–1.79)	(–2.71)	(–3.00)
Herfindahl index of ownership concentration	–0.092*			–0.066		
	(–2.22)			(–1.02)		
Number of municipal owners		0.0195***			0.010*	
		(4.82)			(1.68)	
Use of intermunicipal company or cooperation (=1)			0.096**			0.100*
			(2.62)			(1.89)
Fixed effect for centrality	Yes*	Yes*	Yes**	Yes**	Yes	Yes*
F-values	F = 3.26	F = 2.18	F = 3.33	F = 3.71	F = 2.10	F = 3.00
R-Square	0.24	0.292	0.26	0.41	0.41	0.41
N	268	311	311	175	211	211

Notes: t-values in parentheses, * $p < 0.05$; ** $p < 0.01$; *** $p < 0.001$.

including health care (Hagen, 1997) and education (Borge & Naper, 2005). Similar results have also been obtained in the Swedish case (ESO, 1996). It appears that affluence induces inefficiency, even in services financed by user fees. Other studies on the use of user charges in Norwegian infrastructure sectors suggest that local revenue impacts negatively on total infrastructure fees, including water supply, discharge of sewage, and garbage collection (Borge, 2000). Outside the refuse sector, fees are apparently used as a substitute for ordinary taxes. Since the legal framework requires fees to cover costs only, the impact of revenues is quite similar for costs and user fees. Somewhat surprisingly, shares of population residing in sparsely populated areas have a negative impact of costs and fees. The impact of centrality and settlement patterns are relatively small, which is in line with previous studies of the refuse sector (see review in Dijkgraaf & Gradus, 2003, 153–154).

5.5 Conclusions

Communities across Europe are seeking to consolidate local authorities to improve service delivery and take advantage of economies of scale. At the same time, citizens are unwilling to approve consolidations between neighboring local authorities. People want more and better services, but appear unwilling to accept the organizational repercussions. Intermunicipal companies represent a substitute for local government consolidations. Such public utilities are prevalent in countries where governments are reluctant to outsource public utilities and where municipal restructuring faces intensive popular resistance. Intermunicipal companies are widespread in Belgium, Denmark, Finland, The Netherlands, Norway and Sweden. They operate in infrastructure sectors, such as waste collection and disposal, sewage treatment, water supply, public transportation and electricity distribution. Perhaps the intermunicipal company offers an opportunity to reap economic gains of large-scale operations without imposing full-fledged consolidations?

The issue addressed here is whether the hybrid organization suffers from one or more governance failures. The corporate governance failure suggests that dispersed and indirect ownership weakens incentives to control a company, leading to agency losses and inferior performance. The political economy failure suggests that elected politicians may pursue other goals than efficient service provision. Intermunicipal companies allow elected politicians even less influence. Such companies are therefore expected to have better performance than companies that are owned by a single public authority.

Empirical analyses presented here suggest that fragmented ownership to public induces cost-inefficiency relative to companies owned by a single political authority. In fact, intermunicipal cooperation creates more problems than it solves. In many cases, efficiency losses due to numerous owners are greater than the cost reductions obtained by operating on a larger scale. These results suggest that active politician-owners improve organizational performance, while passive owners bring about management-controlled organizations with lesser efficiency. The management failures described by political economy appears to be less relevant than those identified in corporate governance theory.

Acknowledgments The comments of Professor Øyvind Bøhren and PhD student Stine Ludvigsen are highly appreciated. This chapter was published as Sorensen RJ (2007) Does dispersed public ownership impair efficiency? The case of refuse collection in Norway. Public Administration 85:1045–1058.

References List

Andersen, L. B., & Blegvad, M. (2006). Does ownership matter for the delivery of professionalized public services? Cost-efficiency and effectiveness in private and public dental care for children in Denmark. *Public Administration*, *84*,147–164.

Bartel, A. P., & Harrison, A. E. (1999). *Ownership versus environment: Why are public sector firms inefficient?* NBER Working paper 7043, National Bureau of Economic Research.

Becht, M., Bolton, A., & Roell, A. (2003). Corporate governance and control. In *Handbook of the Economics of Finance*. Oxford: Elsevier.

Borcherding, T., Pommerehne, W. W., & Schneider, F. (1982). Comparing the efficiency of public and private production: A survey of the evidence from five federal states. *Zeitschrift für Nationalœkonomie/Journal of Economic Theory Supplement*, *2*, 127–156.

Borge, L. E. (2000). Charging for public services: The case of utilities in Norwegian local governments. *Regional Science and Urban Economics*, *30*, 703–718.

Borge, L. E., & Naper, L. R. (2005). *Efficiency potential and efficiency variation in Norwegian lower secondary schools*. Working Paper 6205, Department of Economics, Norwegian University of Science and Technology.

Boycko, M., Shleifer, A., & Vishny, R. (1996). A Theory of Privatization. *Economic Journal*, *106*, 309–319.

Bøhren, Ø., & Ødegaard, B. A. (2005). Governance and performance revisited. In G. Gregouriu & P. Ali (Eds.), *International Corporate Governance after Sarbanes-Oxley*. Chichester: Wiley.

Caves D. W., & Christensen L. R. (1980). The relative efficiency of public and private firms in a competitive environment: The case of Canadian railroads. *Journal of Political Economy*, *88*, :958–976.

D'Souza, J., & Megginson, W. L. (1999). The financial and operating performance of privatized firms during the 1990s. *Journal of Finance*, *54*, 1397–1438.

Dixit, A. (1997). Power of incentives in private versus public organizations. *American Economic Review*, *87*,:378–382.

Dijkgraaf, E., & Gradus, R. H. J. M. (2003). Cost savings of contracting out refuse collection. *Empirica*, *30*, 149–161 (reproduced in chapter 2).

Domberger, S., & Jensen, P. (1997). Contracting out by the public sector: Theory, evidence, prospects. *Oxford Review of Economic Policy*, *13*, 67–78.

Dubin, J. A., & Navarro, P. (1988). How markets for impure public goods organize: The case of household refuse collection. *Journal of Law, Economics and Organization*, *4*, 217–241.

Dunsire, A., Hartley, K., Parker, D., & Dimitriou, B. (1988). Organizational status and performance: A conceptual framework for testing public choice theories. *Public Administration*, *66*, 363–388.

Dunsire, A., Hartley, K., & Parker, D. (1991). Organizational status and performance: Summary of the findings. *Public Administration*, *69*, 21–40.

ESO (Expertgruppen för studier i offentlig ekonomi) (1996). *Productivity trends in the public sector in Sweden*. Stockholm: The Ministry of Finance.

Fama, E. F., & Jensen, M. C. (1983). Separation of ownership and control. *Journal of Law and Economics*, *26*, 301–325.

Gomez-Lobo, A., & Szymanski, S. (2001). A law of large numbers: Bidding and compulsory competitive tendering for refuse collection contracts. *Review of Industrial Organization*, *18*, 105–113.

Hagen, T. P. (1997). Agenda setting power and moral hazard in principal-agent relationships: Evidence from hospital budgeting in Norway. *European Journal of Political Research*, *31*, 287–314.

Hall, D. (2006). *Waste management companies in Europe*. Report commissioned by the European Federation of Public Service Unions (EPSU).

Hjalmarsson, L., & Veiderpass, A. (2002). Productivity in Swedish retail electricity distribution. *Scandinavian Journal of Economics*, *94*, 193–205.

Sappington, D. E. M., & Stiglitz, J. E. (1987). Privatization, information and incentives. *Journal of Policy Analysis and Management*, *6*, 567–582.

Savas, E. S. (2000). Privatization and public-private partnerships. New York: Chatham House.

Shirley, M. M. (1999). Bureaucrats in business: The roles of privatization versus corporatization in state-owned enterprise reform. *World Development*, *27*, 115–136.

Shleifer, A., & Vishny, R. W. (1994). Politicians and firms. *Quarterly Journal of Economics*, *109*, 995–1025.

Shleifer, A., & Vishny, R. W. (1997). A survey of corporate governance. *Journal of Finance*, *52*, 737–783.

Sørensen, R. J., & Bay, A. H. (2002). Competitive tendering in the welfare state perceptions and preferences among local politicians. *Scandinavian Political Studies*, *4*, 357–384.

Statistics Norway (2004). Local government-owned enterprises 1999–2004, http://www.ssb.no/english/

Villalonga, B. (2000). Privatization and efficiency: Differentiating ownership effects from political, organizational, and dynamic effects. *Journal of Economic Behavior and Organization*, *42*, 43–74.

Vining, A., & Boardman, A. (1992). Ownership versus competition: Efficiency in public enterprise. *Public Choice*, *73*, 205–239.

Chapter 6
Refuse Collection in Spain: Privatization, Intermunicipal Cooperation, and Concentration

Germà Bel

Abstract This chapter analyzes the structure and dynamics of the refuse collection service in Spain. For this purpose, a database is obtained by means of the II Survey on Local Public Services Delivery. First, results are presented concerning the structure of the refuse collection service. The analysis shows that private production is widely extended in Spain, and hybrid organizational forms such as mixed public-private firms (mixed firms henceforth) have a relevant role. Besides, almost half of the municipalities with population above 2,000 deliver refuse collection through intermunicipal cooperation. Cooperation is more frequent among small municipalities, which indicates that it is used as a formula to take advantage of scale economies. In addition, intermunicipal cooperation is more frequent in municipalities that retain public production, and this suggests that it is being used as an alternative to privatization as well. Finally, concentration is analyzed in the private market for refuse collection, and the analysis shows high degrees of market concentration. Concentration is even more intense at the regional level, and this could pose important obstacles to the competition for the contract.

Keywords Privatization · cooperation · concentration · collection · Spain

6.1 Introduction

Empirical works on the relationship between privatization and costs savings published in the recent years usually show no significant differences between public and private production. Certainly, Reeves and Barrow (2000) find cost savings with privatization in Ireland. However, no significant differences between private and public production costs are found in Callan and Thomas (2001) for Massachusetts -US-, and in Bel (2006a) for Catalonia -Spain-. Ohlsson (2003) finds private production

✉ G. Bel
Universitat de Barcelona and ppre-IREA, Dep. de Política Econòmica,
Av. Diagonal, 690 08034 Barcelona, Spain
e-mail: gbel@ub.edu

more costly in Sweden. On their side, Dijkgraaf and Gradus (2003) find that private firms have lower costs than public bureaucratic production in the Netherlands, but the difference between public and private firms disappears when the service is contracted by means of competitive tendering.[1]

Focusing on the empirical studies for Spain, several works have used Data Envelopment Analysis (DEA) to analyze the technical efficiency of refuse collection, and have compared public and private production. Vilardell i Riera (1989) is the first published study, uses a sample of 46 municipalities of Catalonia, and founds private production more efficient. However, the number of municipalities with public production is too small (6), so that comparing public-private lacks statistical robustness. Still for Catalan municipalities, Bosch, Pedraja, and Suárez-Pandiello, (2000) do not find significant differences in productivity between public and private production with their sample of 75 municipalities. Alvarez Villamaría, Estévez, and González Martínez (2005) DEA analysis for 39 municipalities of Galicia finds that municipalities with public production show more efficient results than those with private production. Finally, Benito, Bastida, and García (2006) study several services in 39 municipalities in the region of Murcia and find that public management of refuse collection is more efficient than private management, but the statistical significance of their results is low. Besides those DEA analysis, there is one multivariate econometric analysis (Bel, 2006a) that finds no significant differences between public and private production with respect to cost in his analysis of 186 municipalities of Catalonia.

Recent and increasing evidence on the lack of costs savings with refuse collection has induced interest in analyzing the dynamics of the management reform in this service, as well as the dynamics of the structure of the private market (Bel, Hebdon & Warner, 2007). In this way, it has been observed that alternative reforms to exploit scale economies through intermunicipal cooperation can reduce one of the advantages that private production was likely to provide initially (Bel & Costas, 2006). In the same direction, several studies (Bel & Costas, 2006; Dijkgraaf and Gradus, 2007) have found that concentration in the private market for refuse collection could be affecting cost savings because of reduced competition.

In this chapter, we analyze the structure and dynamics of the refuse collection service in Spain. Up to now, econometric analyses on the market structure and dynamics of this service in Spain have been limited to data from the region of Catalonia (e.g., Bel and Miralles, 2003, 2004; Bel & Costas, 2006). Here we use a database for the whole of Spain. This database has been obtained by means of the II Survey on Local Public Services Delivery,[2] II Survey hereafter. To our knowledge, the data

[1] Bel & Warner (2007) contains an up-dated literary meta-analysis of the empirical literature on local privatization and costs. Bel, Fageda and Warner (2007) present a meta-regression analysis to find out what explains the divergence in results among the empirical studies on local privatization (solid waste and water) and costs.

[2] The research unit *Public Policies and Economic Regulation*, at the University of Barcelona, developed the *II Survey on Local Public Services Delivery*. The survey focused on the municipalities with population over 2,000 inhabitants. Bel (2006b) displays full information on the *II Survey*.

from this survey allows conducting the first countrywide analysis of market structure and dynamics for refuse collection in Spain.

The first section presents the results obtained concerning the structure of the refuse collection service. The analysis shows that private production is widely extended in Spain, and hybrid organizational forms such as mixed public-private firms (mixed firms henceforth) have a relevant role. As explained in Warner and Bel (2008), in mixed firms ownership is divided between the government and the private sector. Usually, the government retains a control stake in the firm, but the firm operates under private commercial law. The private partner tends to be a large firm with a solid position in the market for private production of the local service. In such cases, day-to-day operations are usually conducted by the industrial private partner, whereas the government retains control over strategic decisions. Under this organizational form, local (or supra-local) governments engage in long term contracts with private firms through joint ventures.

The second section studies intermunicipal cooperation in Spain and finds that almost half of the municipalities with population above 2,000 deliver refuse collection through intermunicipal cooperation. It is important to notice that intermunicipal cooperation in Spain does not involve municipal governments contracting out the service to another government or public agency.[3] Instead, they engage in city partnerships under a joint authority (either a supralocal institution, at county or province level, or a single purpose agency) in governance of which all involved governments play a role. Within this framework, the choice is made between using private production, public production or mixed firms for the service delivery. Hence, cooperation and privatization are not incompatible (Bel & Fageda, 2006, 2008). Cooperation is more frequent among small municipalities, which indicates that it is used as a formula to take advantage of scale economies. In addition, intermunicipal cooperation is more frequent in municipalities that retain public production, and this suggests that it is being used as an alternative to privatization as well.

The third section analyzes concentration in the private market for refuse collection. The analysis shows high degrees of market concentration. Concentration is even more intense at the regional level, and this could pose important obstacles to the competition for the contract. In the last section the main conclusions are drawn.

The Spanish Ministry of Science and Technology (BEC2003-01679) provided financial support, for material and communication expenses. The Rafael del Pino Foundation financed the research assistance during the year 2004–2005. I am thankful to Daniel Albalate, Laia Doménech, Xavier Fageda, David López, Antonio Miralles, and Melania Mur for their assistance and suggestions.

[3] It is worthwhile noting that intermunicipal cooperation in Spain, as well as in many European countries, is compatible with private production. On the contrary, intermunicipal cooperation in other countries is linked to public production, In this way, in Netherlands municipalities that engage in cooperation keep public production (Dijkgraaf and Gradus, 2007). Intermunicipal cooperation differs too from intermunicipal cooperation in the US, since in that country it is seen as a form of contracting out to another government or public agency (Warner & Hefetz, 2002a, b).

6.2 The Structure of the Solid Waste Service Production in Spain

The development of the survey has permitted to obtain useful information about the form of production of refuse collection in 540 municipalities. The sample includes 25% of the Spanish municipalities with population over 2,000 inhabitants. All municipalities with population over 30,000 inhabitants are included. If we take into account the population included in the sample, the representativity is much higher than the one we found when considering the number of municipalities. The population included in the sample represents almost 75% of the aggregated population in the cities/towns with more than 2,000 inhabitants, and almost 70% of the total population of Spain. This is explained because the frequency of answers to the survey increases with population size.

The current level of contracting out in the urban solid waste service is very high. Table 6.1 presents the shares of each production form in Spain. Results are presented for different municipality sizes (population), as well as for the whole set of Spanish municipalities with population over 2,000.[4]

More than half (56%) of the Spanish municipalities with more than 2,000 inhabitants choose private production (equivalent to contracting out in practice in Spain), while 37% use public production, be it done either directly by the bureaucracy or by means of a public firm operating under private commercial law. In the remaining municipalities, 7%, governments keep some degree of control on the delivery of the

Table 6.1 Form of production of refuse collection in Spain 2003: municipalities

Municipality size (population)	Public Unit (bureaucracy)	Public firm	Mixed firm	Private firm	Other
2,001–10,000	27.1%	13.0%	8.2%	51.7%	0.0%
10,001–30,000	20.0%	9.2%	4.3%	66.5%	0.0%
30,001–50,000	14.5%	18.4%	2.6%	64.5%	0.0%
50,001–100,000	9.6%	11.0%	5.5%	72.6%	1.4%
More than 100,000	14.0%	17.5%	3.5%	64.9%	0.0%
Total (weighted)	24.2%	12.4%	7.0%	56.3%	0.1%

Notes: In building the figure for 'Total', gross percentages resulting from the sample have been adjusted. To do so, we have taken into account the different frequency of responses to the Survey obtained across municipality sizes. *Other* means that the city territory is divided in different service areas. One (or some) of them are served by public firms and one (or some) are served by private firms.
Source: Author.

[4] These categories are as follows: a) Public units or public agencies that operate within the framework of the administrative public law; b) Public firms that operate within the framework of private commercial law; c) Firms that are partially owned by the government and partially owned by the private sector; and d) private firms. It is worth noting that, different to countries like the Netherlands, the US or Norway, in the case of Spain it is exceptional for a contract to be won by a public firm of a different municipality. To that effect, in Spain contracting out and private production are equivalent in practice.

service,which is done by mixed firms (firms with partial government ownership and partial private ownership). In these last cases, we cannot refer strictly to privatization. However, this is a formula for having private sector participation in the delivery of the service. Therefore, private participation (considered in a more generic meaning than that of just pure private production) reaches 63% of the municipalities.

Within the context of public production, direct (bureaucratic) production is more frequent in the municipalities with less population, while the using of public firms is more frequent in municipalities with more population. An interesting exception to this general pattern is the relatively high presence of public firms in the smallest municipalities, especially those between 2,000 and 5,000 inhabitants.[5] It is worth noting that many small municipalities have engaged in intermunicipal cooperation, as we will extensively analyze below, and public firms operating in these municipalities are usually companies of county or regional dimension. On the contrary, public firms of strictly municipal dimension are very scarce in this population range.

The differences between public production and private production are more important if we consider the population involved, as shown by results in Table 6.2. The average population of the municipalities with public production tends to be lower than the average population of the municipalities with private production. However, the wide extension of contracting out has implied that many municipalities with small population have adopted private production: indeed, more than half of the municipalities with population between 2,000 and 10,000 inhabitants use private delivery.[6]

We have found privatization to increase with population, up to a maximum frequency in the intermediate municipalities. This is consistent with findings in Bel and Miralles (2003) and the interpretation is similar. Municipalities with relatively small population, and thus small demand for refuse collection service, can take little quantitative benefit from reducing inefficiencies through privatization, while they

Table 6.2 Form of production of refuse collection in Spain 2003: population

Municipality size (population)	Public unit (bureaucracy)	Public firm	Mixed firm	Private firm	Other
2,001–10,000	26.1%	10.8%	8.8%	54.3%	0.0%
10,001–30,000	19.6%	9.5%	4.5%	66.4%	0.0%
30,001–50,000	15.1%	18.9%	2.6%	63.4%	0.0%
50,001–100,000	9.5%	10.8%	5.0%	73.0%	1.7%
More than 100,000	9.5%	14.1%	5.3%	71.1%	0.0%
Total (weighted)	14.8%	12.4%	5.6%	67.0%	0.2%

Note: See notes in Table 6.1.
Source: Author.

[5] The frequency of the public firm is 15.6% in the municipalities of population between 2,001 and 5,000 inhabitants, but is only 8.5% in municipalities between 5,001 and 10,000 inhabitants.

[6] Private production in the municipalities with population below 5,000 is 46.8%. The share of private production is identical to that of the sum of the public production forms (direct administration + public firm = 46.7%).

face relatively large transaction costs in the case they contract out. The implementation of the tendering process and the supervision of the contracted service imply a type of fixed cost that can reach large relevance with respect to total cost for the service for those municipalities with small population. These factors help to explain the small frequency of contracting out in this type of municipalities. Overall, more moderate frequency of private production in small municipalities is likely reflecting the fact that transaction costs are relatively more important in these towns.[7]

As population increases, however, transaction costs decrease relatively, and the expected benefits from contracting out increase; this makes the cities more prone to contract out. However, economies of scale are exhausted after some city size is reached. Hence, frequency of contracting out stops increasing after certain population level.

It is worth noting that the frequency of public production is relatively higher in the smallest municipalities in Spain, contrarily to what happens in other countries for which there exists enough information, such as Italy, Sweden and the Netherlands (Bel, 2006b). No direct explanation for this fact can be provided. According to the hypothesis in Bel and Miralles (2003) on relatively higher transaction costs under privatization in small municipalities, relevance must be given to the fact that average population of the Spanish municipalities is much lower than average population of municipalities in Italy, Sweden and the Netherlands. This is so because the Spanish municipal map is much more fragmented, and there exist many more municipalities with low population. Instead, the smallest urban places in Sweden, Italy and the Netherlands are integrated in wider municipalities, thus more prone to contracting out.[8]

6.3 Intermunicipal Cooperation

Many municipalities have a size smaller than the optimal for the delivery of local public services. Of course, the optimal size differs depending on the type of service. In any case, local privatization is a formula that can be used to exploit scale economies (Donahue, 1989). Contracting out allows a private firm to deliver the service in different municipalities within the same geographical area, and this

[7] Other additional factors can help to explain why small municipalities have less privatization. Among them, Warner (2006), and Warner and Hefetz (2002b, 2003) have stressed lower managerial capabilities to deal with private firms, as well as lower attractiveness of small municipalities for those private firms offering delivery of services.

[8] Indeed, average municipality size (population) is much higher in all those countries than in Spain. Data for population and municipal structure for 2004 show that average municipal population in Spain was 5,327 inhabitants. Instead, Italy -with roughly the same number of municipalities than Spain has (8,010 according to the 2001 Census)- had 57.3 million people in 2004, which means that average municipal population is 7,076 inhabitants, more than 30% higher than the Spanish average. The difference between Spain and Sweden and the Netherlands is much larger: in 2004 average municipal population was 35,000 inhabitants in the Netherlands and almost 31,000 inhabitants in Sweden.

would permit the private firm to benefit from scale economies in its production costs. However, privatization is not the unique formula that can be used to exploit scale economies; the aggregation of the service across municipalities by means of intermunicipal cooperation permits too exploiting scale economies. Delegating the governance of the service (from the municipality) to a joint authority (e.g. province, county or special agency) is compatible with public production as well as with private production.

Intermunicipal cooperation for the production of the service is widely extended in Spain. The decision of municipal governments to engage in intermunicipal cooperation is discretional.[9] What does the information obtained by means of the survey say about the existence of intermunicipal cooperation in the solid waste collection service? Table 6.3 presents synthetic information on the frequency of intermunicipal cooperation. The information is presented for Spain as a whole, as well as for those *Comunidades Autónomas* (Spanish regions consisting of one or more provinces) for which the level of response to the survey allows having a moderate sample error in statistical terms.

In Spain as a whole, almost half of the municipalities with population above 2,000 inhabitants deliver refuse collection service through intermunicipal cooperation.[10] However, there exist large differences between territories concerning the frequency of intermunicipal cooperation. For example, more than 70% of the municipalities in Andalusia cooperate in this service, while aggregation does not reach 15% in the Comunidad Valenciana and in Madrid.

A common pattern, which is observed without any exception, is the decrease in the frequency of intermunicipal cooperation as population increases. This happens all over Spain, and also in all the regions for which we have sensible data in statistical terms. Empirical studies on costs suggest that the refuse collection service is subject to scale economies[11] and that the optimal size for this service would be within the range of 20,000–50,000 inhabitants. The information displayed

Table 6.3 Intermunicipal Cooperation in refuse collection (in percentage of municipalities)

	N	All	> 30, 000 inh.	10,001–30,000 inh.	2,001–10,000 inh.
Spain	540	44.0	14.6	30.8	52.2
Andalusia	94	70.7	25.6	56.1	78.6
Catalonia	109	34.7	2.9	17.6	50.0
C. Valenciana	65	14.7	4.3	8.1	20.0
Madrid	36	12.2	0.0	0.0	20.0

[9] Martínez Lacambra (2003) analyzes the 'Mancomunidades' (organizations of voluntary adscription) as an associative model for the management of public services, and gives recent data about its proliferation in Spain.

[10] The intermunicipal cooperation can exist in all or in some parts of the service. Cooperation is very frequent in the disposal in landfills or incinerators. When intermunicipal cooperation in collection and transport exists, it generally comprises disposal as well as recycling.

[11] See Stevens (1978), Tickner and McDavid (1986), Dubin and Navarro (1988), Callan and Thomas (2001), Bel & Costas (2006) and Dijkgraaf and Gradus (2007).

in Table 6.2 is consistent with this observation, since the frequency of intermunicipal cooperation declines in municipalities with population over 10,000, and the decline is especially acute in cities over 30,000 inhabitants.

6.3.1 Intermunicipal Cooperation and Privatization

The fact that the production of the service can be aggregated through intermunicipal cooperation does not restrict, beforehand, the choice concerning the form of production. In fact, production form choices made by municipalities that engage in cooperation are diverse (below is provided a detailed analysis). Governments that have engaged in cooperation can choose to organize the delivery by means of public production, either bureaucracy or public firm, with a mixed firm, or with a private firm.

However, among the likely effects derived from intermunicipal cooperation there is one of special relevance for our purposes: since intermunicipal cooperation allows the exploitation of scale economies no matter the form of production used, it overcomes one of the most important advantages that private production could offer to governments that provide refuse collection at the municipal level. This is specially so in the case of the smallest municipalities. With regard to this type of municipalities, Warner and Hefetz (2003) suggest that the intermunicipal cooperation can be an extremely operational alternative to privatization, since private firms are less attracted to work in this type of municipalities given the small scale of operations involved.

Does there exist any clear relationship between intermunicipal cooperation and privatization? In Table 6.4 we divide the sample between municipalities that cooperate and municipalities that do not cooperate. This allows comparing the frequencies of each form of production in all types of municipalities.

According to the results obtained those municipalities that have engaged in intermunicipal cooperation show a much lower frequency of private production than those municipalities that keep service provision at the municipal level (no cooperation). Contrarily, pure public production (public firm + public bureaucracy) is much more frequent in the municipalities engaged in cooperation than in those that do not cooperate. In addition, mixed firm is much more frequent among the municipalities that cooperate (10.1%) than among those that do not (2.4%).

The results clearly suggest that many municipalities use intermunicipal cooperation to take advantage of scale economies. At the same time, cooperation can be used to gain bargaining power (as relative to the limited power each municipality has

Table 6.4 Intermunicipal Cooperation in refuse collection (percentage)

Form of production	Supramunicipal or intermunicipal	Municipal
Private	51.3	71.5
Mixed	10.1	2.4
Public Firm	16.5	10.2
Public Bureaucracy	22.2	15.7
Other	0.0	0.3

taken as single entity) in order to find an industrial private partner that could help to improve the service delivery. This could explain the higher frequency of mixed firm in municipalities that cooperate; instead, in the case of municipal provision the choice seems much more drastic between pure public production and pure private production.

To conclude, intermunicipal cooperation is compatible with all forms of production. However, our results show that privatization in the municipalities that cooperate is less frequent than in those that do not.[12] This result is consistent with the hypothesis suggested above: intermunicipal cooperation overcomes one of the main economic advantages provided by privatization, its using as a formula to take advantage of scale economies.

6.4 Concentration and Competition in the Private Market for Concessions

The theoretical and empirical literature on local privatization suggests that, from the point of view of the efficiency, the degree of competition *for* the sector is more important than the ownership structure of the firm (Bel & Warner, 2007). Indeed, the potential effects from ownership largely depend on the intensity of the incentives derived from competition. Within this context, the existence of problems related to competition for the contracts could be an important limitation to the potential improvements achieved through private production. This could likely affect technical efficiency, which would be reflected in the firm's production costs; and this could affect allocative efficiency too, which would be reflected on the costs paid for the service by municipalities or by users. For these reasons, it is interesting and useful analyzing the degree of concentration in the private market of solid waste collection.

The *II Survey* has allowed obtaining the identification of the firm contracted for the concession in 346 of the 354 municipalities with private production of refuse collection. In Spain the private market for concessions in refuse collection is practically equivalent to the total market for concessions, since public firms usually do not bid for contracts in other municipalities outside their hometown.[13] The data collected allow us knowing and analyzing the distribution of concessions by firms. A specific analysis of the degree of concentration in the market can be done by using concentration indexes well-established in the economic literature. We use here the concentration ratios CR1 (quota of the leading firm in the market), CR4 (quota of the four main firms in the market) and HHI (Herfindahl-Hirschman index).

[12] Bel and Fageda (2008) empirically document such negative relationship.

[13] We have been able to identify only one case of municipality that awarded the concession to a public firm belonging to a different municipality. This case refers to a medium-size municipality in the region of Madrid. Obviously, results would be almost exactly the same if we considered this exceptional case in the computation of the concentration indexes. Hence, there is no need to include a concentration rate for public and private firms, as it is actually done in Dijkgraaf and Gradus (2007), since this would not make sense for Spain.

- Concentration rate is defined as the cumulative market share of the first k firms:

$$CR_k = \sum_{i=1}^{k} X_i / X = \sum_{i=1}^{k} Q_i,$$

where X_i is the *ith*-firm's size, X is the total market size, and Q_i is the *ith*-firm's market share. This rate can be either calculated for the leading firm (CR_1), or it can be taken as a 'marginal concentration rate', combining a given number of firms. In the latter case, the most usual indicator used in the literature is the one that includes the first four firms (CR_4).

- Hirschman-Herfindahl Index (HH) is defined as the sum of squared market shares:

$$HH = \sum_{i=1}^{n} Q_i^2,$$

where Q_i is defined as above and n is the total number of firms in the market. Its advantage over simple concentration rates consists in its ability to take into account both the number of firms and the differences among them, as big firms are weighted high and small firms weighted low. Thus, as pointed out by Yoo (2002), the Hirschman-Herfindahl index has become the standard concentration measure for antitrust enforcement purposes.

6.4.1 Concentration and Competition in Waste Collection

The total number of concessions in the sample is of 350, because two municipalities, Barcelona (four concessions) and Valencia (two concessions),[14] have more than one private firm delivering the service. Table 6.5 presents the results derived from the *II Survey* about the structure of the concessions market in Spain. The firms whose market share is higher than 1% of the concessions are identified. Those firms with market shares smaller than 1% are aggregated in two different subsets: (1) firms whose market share is between 0.5% and 1%; and (2) firms with a market share below 0.5%.

The degree of concentration is very high. This happens with respect to the market shares for concessions and, more intensely, for the market shares concerning the

[14] Since 2000 Barcelona is divided in four areas for the service of refuse collection and street cleaning, and correspondingly the concession contracts awarded are four. To stimulate a competitive process and to make benchmarking easier, the number of concessions that a particular firm can win is limited to a maximum of two (50%). The firms that obtained concessions are FCC (2), Ferrovial-CESPA and Urbaser. In the city of Valencia, since 1994 the service is divided in two concessions: FCC is in charge of the Northern area of the city, and Sociedad Agricultores de la Vega is in charge of the Southern area.

Table 6.5 Market structure of the concessions 2003

Firm	Share of concessions (%)	Share of population with private delivery (%)	Average population by concession
Fomento de Construcciones y Contratas	32.9	52.0	92,924
Ferrovial /CESPA	18.3	16.9	54,148
ACS/Urbaser	14.0	16.2	68,034
Acciona/Necso	2.9	1.2	23,957
Fomento Benicásim (FOBESA)	1.4	0.6	24,990
Sociedad Agricultores de la Vega	1.4	2.2	91,874
Canaria de Limpiezas Urbanas (CLUSA)	1.1	0.5	25,575
Firms with market share 0.5-1.0%	8.3	3.5	24,525
Firms with market share lower than 0.5%	19.7	7.0	20,804
Total			58,761

population served by private firms. A single company, Fomento de Construcciones y Contratas (FCC), concentrates more than 30% of the concessions and more than 50% of the population served by private enterprises. After FCC, we find Ferrovial-CESPA, the second largest group in the sector, which concentrates 18% of the concessions and 17% of the served population. The third group, ACS-Urbaser, has 14% of the concessions and serves 16% of the population. The remaining groups and individual firms have much smaller shares: only four of them present a market share of concessions higher than 1%, and three of them are firms that operate just in a single region.

On the other hand, more than a quarter of the concessions (28%) have been awarded to firms of very small dimension that operate only in specific territorial areas. This shapes a dual market. On one side, at the upper end of the market, there is a high degree of concentration due to the accumulation of contracts in the biggest companies. On the contrary, at the lowest end of the market, there exists an intense degree of fragmentation, concentrated in the smallest municipalities. In fact, more than half of the firms that have lowest market shares serve municipalities which population is below 10,000 inhabitants.

The market share of concessions computed so far is biased upward for the biggest companies, given that the large firms are particularly prone to have concessions in the large cities, and large cities are overrepresented in the sample. The upward bias for the large firms is not important when calculating market shares with respect to the population served, since representativity of the sample is extremely high with regard to population. Even if there could exist a slight upward bias of the large firms' market share, its effect on the results obtained is likely to be marginal.

Table 6.6 presents the information on the main firms and on the concentration indexes computed for the number of concessions. First, estimates are presented for the whole set of municipalities. These estimates have an upward bias for big firms, as mentioned above. Second, we present computations for municipalities with population above 10,000; the potential bias for these estimates is much lower, since

Table 6.6 Market shares (in %) and concentration indexes: concessions

	N	Leading		Second		Third		CR1	CR4	HHI
		Group	Share	Group	Share	Group	Share			
Total	350	FCC	32.9	Ferrovial	18.3	Urbaser	14.0	0.33	0.68	0.16
Pop > 10,000	282	FCC	36.2	Ferrovial	19.1	Urbaser	16.7	0.36	0.75	0.20
Pop > 30,000	144	FCC	39.9	Ferrovial	22.9	Urbaser	19.6	0.40	0.84	0.25
Pop. 10,001–30,000	138	FCC	32.6	Ferrovial	15.1	Urbaser	13.0	0.33	0.64	0.15
Pop < 10,000	68	FCC	19.1	Ferrovial	14.7	*	*	0.19	0.41	0.07

Note: * The sample of municipalities with population below 10,000 does not allow to identify with enough accuracy the name of the third group/firm in this segment.

under-representation in the sample focuses in the 2,000–10,000 segment. Finally, computations are presented for three different segments according to the population of the municipality; here the potential bias, when existing,[15] is only marginal.

Table 6.7 presents the same type of computations, with the same level of disintegration, but this time taking into account the population served by private firms. FCC is the leading group in the sector, and this is so for the number of concessions as well as for the population served. FCC holds the leading position for the whole sample as well as for all segments according to the population size that we have established. The only exception we find is in the segment of municipalities with population below 10,000 inhabitants; here, although FCC is still the group with higher market share concerning the number of concessions, Ferrovial is the group with higher market share when taking into account the population served. This last group, Ferrovial, is precisely the second one in the market in almost all the segmentations done with regard to population. The unique exception to this pattern is that of Urbaser placed as second group in the segment of intermediate municipalities (with population between 10,001 and 30,000 inhabitants) when taking into account the population served.

The ratios CR1 and CR4 show a very high level of concentration. The market share of the leading firm reaches 30% of concessions and 50% of the population served, as mentioned above. If we consider separately the market of municipalities of more population (pop. >30,000), the shares of the leading firm are even higher.

Table 6.7 Market shares and concentration indexes: population served by private firms

	N	Leading		Second		Third		CR1	CR4	HHI
		Group	Share	Group	Share	Group	Share			
Total	350	FCC	52.0	Ferrovial	16.9	Urbaser	16.2	0.52	0.86	0.33
Pop > 30,000	144	FCC	55.2	Ferrovial	17.1	Urbaser	16.1	0.55	0.91	0.36
Pop. 10,001–30,000	138	FCC	33.4	Urbaser	16.2	Ferrovial	14.7	0.33	0.68	0.16
Pop < 10,000	68	Ferrovial	19.7	FCC	17.7	*	*	0.20	0.46	0.08

Note: * The sample of municipalities with population below 10,000 does not allow to identify with enough accuracy the name of the third group/firm in this segment.

[15] No bias exist for the >30,000 segment, since the sample includes all cities of this size.

When adding the market shares of the four biggest groups we obtain that these firms concentrate 2/3 of the concessions (3/4 in the municipalities of medium and large size) and almost 90% of the population.

If we refer to the four types of market structures characterized in Buesa and Molero (1998) for the analysis of the CR4, the level of concentration is very high (level 1, above 60%) for concessions, as well as for population served. Only in the smallest municipalities we find levels of concentration of type 2 (high), although we have seen that the smallest firms have a very small size and very limited territorial operation, what suggests a scarce potential for competition.

The Herfindahl-Hirschman index is close to 0.2 for concessions and is higher than 0.30 for the population. According to the four types of market structure characterized in Besanko, Dranove, and Shanley (2000), the structure is of monopolistic competition (here competition would require product differentiation) with respect to concessions. Regarding the population, we find a structure of oligopoly. In view of the limits existing in this sector for the differentiation of products, the market structures we have found suggest that competition is scarce, and it depends crucially on the rivalry among firms.

The fact that our analysis has focused on national market so far does not necessarily mean that national market is the relevant market. In fact, Dijkgraaf and Gradus (2007) suggest that smaller relevant markets should be considered particularly for small and medium-size municipalities. Hence, it is worth providing a more detailed analysis of concentration levels in the Spanish regions.

Tables 6.8 and 6.9 display the analysis on concentration carried out for the regional level. Our analysis is more robust for those regions for which our sample includes 10 or more observations of municipalities with private delivery. In fact, those regions for which our sample of municipalities with private delivery is below 10 do not allow for a sensible computation of concentration indexes. Still, concentration indexes are provided for informative purpose. When analyzing the population served (Table 6.9) statistical significance of our computations increases, since representativeness with respect to population is much higher than with respect to concessions.

As a common pattern, the results at the regional level show degrees of concentration that are higher than those found for Spain as a whole. Focusing of concentration with respect to population, only in the case of Balearic Islands, Valencia, Castile & Leon, Galicia, and Andalusia the HH index reflects a structure of monopolistic competition. In the other regions, the market structure is of oligopoly, with a clear position of control by one single firm in the cases of Madrid, Catalonia and Canary Islands, as well as all the smaller regions (bottom rows in the table) with the exception of Balearic Islands and Extremadura.

The case of Andalusia is rather singular: on one hand, there exists a very high degree of concentration regarding the number of concessions. Nonetheless, the fact that the shares with respect to population are more divided than in other regions generates a situation of higher potential for competition, as compared to other regions. Another interesting result is that of the two archipelagos: Canary Islands and Balearic Islands have important local/regional providers with strong leading

Table 6.8 Market shares (in %) and concentration indexes: concessions

	N	Leading		Second		Third		CR1	CR4	HHI
		Group	Share	Group	Share	Group	Share			
Andalusia	36	Urbaser	33.3	Ferrovial	27.8	FCC	22.2	0.33	0.86	0.24
Catalonia	98	FCC	37.8	Ferrovial	16.3	Urbaser	8.2	0.38	0.67	0.18
C. Valenciana	58	FCC	37.9	SAV	8.6	Fobesa	8.6	0.38	0.64	0.18
Madrid	20	FCC	40.0	Ferrovial	40.0	Urbaser	15.0	0.40	1.00	0.35
Murcia	10	Ferrovial	30.0	FCC	30.0	Urbaser	30.0	0.30	1.00	0.28
Canary Islands	15	Urbaser	26.7	CLUSA	26.7	**	**	0.27	0.67	0.17
Basque Country	19	Ferrovial	57.9	FCC	31.6	**	**	0.58	1.00	0.44
Castile & Leon	19	FCC	26.3	Urbaser	21.1	**	**	0.26	0.68	0.16
Castile laMancha	14	Ferrovial	50.0	FCC	21.4	FOMA	14.3	0.50	0.86	0.33
Galicia	21	Urbaser	38.1	Ferrovial	14.3	FCC	9.5	0.38	0.71	0.20
Extremadura	8	FCC	50.0	Coniser	25.0	*	*	0.50	**	0.34
Asturias	2	FCC	50.0	*	*	*	*	0.50	**	0.50
Aragon	6	FCC	66.7	*	*	*	*	0.67	**	0.50
Balearic Islands	9	LUMSA	25.0	Ferrovial	25.0	*	*	0.25	0.75	0.19
Navarre	4	FCC	100.0	*	*	*	*	1.00	0.00	1.00
Cantabria	5	SADISA	40.0	*	*	*	*	0.40	**	0.28
La Rioja	4	FCC	75.0	Urbaser	25.0	*	*	0.75	**	0.63

Notes: * The sample of municipalities in the region does not allow a sensible computation of concentration indexes. **The sample of municipalities in the region does not allow identifying with enough accuracy the name of the third group/firm. It is important to note that sample in the regions in the lowest part of the table are too small to allow deriving any robust result on concentration indexes. The information for Extremadura, Asturias, Aragón, Balearic Islands, Navarre, Cantabria and La Rioja has only informative purposes.

Table 6.9 Market shares and concentration indexes: population served by private firms

	N	Leading		Second		Third		CR1	CR4	HHI
		Group	Share	Group	Share	Group	Share			
Andalusia	36	Urbaser	41.2	FCC	26.2	Ferrovial	19.8	0.41	0.88	0.28
Catalonia	98	FCC	56.9	Ferrovial	15.3	Urbaser	13.4	0.57	0.90	0.37
C. Valenciana	58	FCC	41.8	SAV	16.1	Ferrovial	12.2	0.42	0.78	0.23
Madrid	20	FCC	84.5	Ferrovial	8.0	Urbaser	7.3	0.85	1.00	0.73
Murcia	10	Ferrovial	59.7	FCC	30.4	Urbaser	8.4	0.60	1.00	0.46
Canary Islands	15	Urbaser	55.2	CLUSA	14.6	**	**	0.55	0.90	0.35
Basque Country	19	FCC	51.6	Ferrovial	47.2	**	**	0.52	1.00	0.49
Castile & León	19	FCC	37.0	Urbaser	30.4	**	**	0.37	0.93	0.27
Castile la Mancha	14	Ferrovial	46.3	FCC	37.8	Urbaser	14.2	0.46	0.99	0.38
Galicia	21	Urbaser	38.0	Ferrovial	28.5	FCC	24.2	0.38	0.93	0.29
Extremadura	8	FCC	51.2	Coniser	37.6	*	*	0.51	**	0.41
Asturias	2	FCC	97.1	*	*	*	*	0.97	**	0.94
Aragon	6	FCC	98.4	*	*	*	*	0.98	**	0.97
Balearic Islands	9	LUMSA	29.5	Ferrovial	28.4	*	*	0.30	0.89	0.22
Navarre	4	FCC	100.0	*	*	*	*	1.00	0.00	1.00
Cantabria	5	SADISA	83.3	Urbaser	8.19	*	*	0.83	**	0.71
La Rioja	4	Urbaser	82.2	FCC	17.8	*	*	0.82	**	0.71

Note: See notes Table 6.8.

positions in their markets. Even if they are not the unique cases of local providers in strong position (see Cantabria and Extremadura), they are far the clearest ones. Maybe geographic isolation (both archipelagos integrate several islands) helps local providers to retain high market quotas.

All in all, the regional detail confirms what we have previously found with the analysis for countrywide Spain. The levels of market concentration are very high, the market structure is not prone to competition for the contracts, and competition depends crucially on the existence of open rivalry between the firms, specially the three largest ones: FCC, Ferrovial, and Urbaser.

6.5 Conclusions

Privatization of solid waste collection is widespread in Spain, since almost 60% of the Spanish municipalities use private firms to deliver the service. Private production share is high in all municipalities' size. Nonetheless, the smallest towns have less private production than the average. Average population of municipalities with private production is larger than that of municipalities with public delivery. Hence, private production share in terms of population is even more hegemonic. Overall, private participation in the delivery (private firms + mixed public-private firms) reaches almost 2/3 of the municipalities and more than 70% of total population.

Privatization increases with population, up to a maximum frequency in the intermediate municipalities. Overall, more moderate frequency of private production in small municipalities is likely reflecting the fact that transaction costs are relatively more important in these towns. As population increases, however, transaction costs decrease relatively, and the expected benefits from contracting out increase; this makes the cities more prone to contract out. However, economies of scale are exhausted after some city size is reached. Hence, frequency of contracting out stops increasing after certain population level.

Some other interesting questions related to local privatization and local public sector reform in Spain have been analyzed: (1) the existence of other alternative reforms, such as intermunicipal cooperation; and (2) the dynamics of the market for local public services, with regards to concentration and competition.

Almost half the municipalities over 2,000 inhabitants cooperate with other municipalities in the delivery of solid waste collection. Frequency of intermunicipal cooperation widely diverges among Spanish regions. However, a common pattern all over Spain is that cooperation is more frequent among small towns and cities. This clearly suggests that intermunicipal cooperation is used in order to exploit scale economies in the delivery. Thus, it is less important among those larger municipalities, since these already operated at optimal scale.

On the other side, intermunicipal cooperation, even if not incompatible with private production, is positively related to public production. Frequency of private production in municipalities that cooperate is lower. On the contrary, pure public production (bureaucratic delivery + municipal public firms) is more frequent among municipalities that cooperate. Hence, this suggests that many municipalities use intermunicipal cooperation to exploit scale economies instead of using privatization. In addition, intermunicipal cooperation can be used to attain a stronger bargaining position in order to undertake cooperation with private firms.

The last section of the chapter has analyzed the structure of the private market for refuse collection services in Spain. Special attention has been paid to the analysis of market concentration, one of the main factors with regard to the prospects for competition for the contract. The concentration degrees, measured with indicators of common use in the analysis of market structures, show that the refuse collection service moves between monopolistic competition and oligopoly. The analysis of the market structures at the regional level shows generally even higher levels of market concentration. Since the data available for the Spanish market relates only to one year, we can not infer conclusions about the dynamics of concentration over time. Comparable data available for the region Catalonia in 2000 and 2006 show a trend of increasing concentration in this region (see Appendix). Future research will devote efforts to gather more recent data for Spain so that dynamics of concentration can be analyzed more in deep.

All in all, the results are discouraging for competition, and weak competition heavily undermines the likelihood of cost savings through privatization. Antitrust authorities should be aware of the problems posed by concentration in the private market for local services, and they should include it into their agendas.

Acknowledgments Research on privatization of local services and the market for refuse collection in Spain has received financial support from the Spanish Commission of Science and Technology (SEJ2006-04985) and from the Fundación Rafael del Pino. I am thankful for many comments and suggestions received from Elbert Dijkgraaf, Xavier Fageda, Raymond Gradus and Mildred Warner.

Appendix

Table 6.10 Change overtime in the concentration index (HHI) in Catalonia

	Data year	HHI concessions	HHI population
Sample 1 (103)	2000	0.106	0.268
Sample 2 (103)	2006	0.131	0.304

Note: In brackets, number of municipalities in the sample. Original sources are the survey used in Bel & Costas (2006) for year 2000 and a more recent survey realized on behalf of the Antitrust Commission of Catalonia for year 2006. Both surveys provided information on a larger sample of municipalities with private production: 152 for 2000 and 200 for 2006. To compare concentration indexes in both years we have selected all municipalities (103) for which information was obtained in both surveys. Hence, the sub-samples used are strictly comparable. Source: Surveys done in 2000 and 2006/2007.

References List

Alvarez Villamaría, X. C., Estévez, M. J. C., & González Martínez, X. M. (2005). Evaluación económica del servicio de recogida de basuras en los municipios gallegos. *Revista de Estudios Regionales*, *72*, 85–112.

Bel, G. (2006a) Un análisis de los gastos municipales por el servicio de residuos sólidos urbanos. *Revista de Economía Aplicada*, *14*, 5–32.

Bel, G. (2006b) *Economía y política de la privatización local.* Madrid: Marcial Pons.

Bel, G, & Costas, A. (2006). Do public sector reforms get rusty? Local privatization in Spain. *Journal of Policy Reform*, *9*, 1–24.

Bel, G., & Fageda, X. (2006). Between privatization and intermunicipal cooperation: Small municipalities, scale economies and transaction costs. *Urban Public Economics Review*, *6*, 13–31.

Bel, G., & Fageda, X. (2008). Reforming the local public sector: Economics and politics in privatization of water and solid waste. *Journal of Economic Policy Reform*, 11, 45–65.

Bel, G., Fageda, X., & Warner, M. (2007) *Privatization and costs: a meta-regression analysis for local services.* Working paper, Universitat de Barcelona & Cornell University.

Bel, G., Hebdon, R., & Warner, M. (2007) Local Government Reform: Privatisation and Its Alternatives. *Local Government Studies*, *33*, 507–515.

Bel, G., & Miralles, A. (2003). Factors influencing privatization of urban solid waste collection in Spain. *Urban Studies*, *40*, 1323–1334.

Bel, G., & Miralles, A. (2004). La competencia en el sector de los servicios públicos locales en España. In L. Cases (Ed.), *Anuario de la Competencia 2003* (pp. 207–222). Madrid: Marcial Pons.

Bel, G., & Warner, M. (2007). *Privatization of solid waste and water services: What happened to costs savings?* Working paper, Universitat de Barcelona & Cornell University.

Benito, B., Bastida, F., & García, J. A. (2006). *The determinants of efficiency in municipal governments.* Working paper University of Murcia and Technical University of Cartagena.

Besanko, D., Dranove, D., & Shanley, M. (2000). *The economics of strategy.* New York: John Wiley & Sons.

Bosch, N., Pedraja, F., & Suárez-Pandiello, J. (2000). Measuring the efficiency of Spanish municipal refuse collections services. *Local Government Studies*, *26*, 71–90.

Buesa, M., & Molero, J. (1998). *Economía Industrial de España.* Madrid: Civitas.

Callan, S. J., & Thomas, J. M. (2001). Economies of scale and scope: A cost analysis of municipal solid waste services. *Land Economics*, *77*, 548–560.

Dijkgraaf, E., & Gradus, R. H. J. M. (2003). Cost savings of contracting out refuse collection. *Empirica*, *30*, 149–161 (reproduced in chapter 2).

Dijkgraaf, E., & Gradus R. H. J. M. (2007). Collusion in the Dutch waste collection market. *Local Government Studies*, *33*, 573–588.

Donahue, J. D. (1989). *The privatization decision. Public ends, private means.* New York: Basic Books.

Dubin, J., & Navarro, P. (1988). How markets for impure public goods organize: the case of household refuse collection. *Journal of Law, Economics & Organization*, *4*, 217–241.

Martínez Lacambra, A. (2003). *Las mancomunidades como modelo asociativo para la gestión de servicios públicos.* Barcelona: Àrea Metropolitana de Barcelona.

Ohlsson, H. (2003). Ownership and production costs. Choosing between public production and contracting-out in the case of Swedish refuse collection. *Fiscal Studies*, *24*, 451–476 (reproduced in chapter 4).

Reeves, E., & Barrow, M. (2000). The impact of contracting-out on the costs of refuse collection services.The case of Ireland. *Economic and Social Review*, *31*, 129–150.

Stevens, B. J. (1978). Scale, market structure, and the cost of refuse collection. *Review of Economics and Statistics*, *60*, 438–448.

Tickner, G., & McDavid, J. C. (1986). Effects of scale and market structure on the costs of residential solid waste collection in Canadian cities. *Public Finance Quarterly*, *14*, 371–393.

Vilardell i Riera, I. (1989). L'eficiència en l'actuació de les administracions municipals: Una avaluació del servei de recollida de residus sólids urbans. *Revista Económica de Catalunya, 11*, 20–33.

Warner, M. (2006). Market-based governance and the challenge for rural governments: U.S. trends. Social Policy and Administration: *An International Journal of Policy Research, 40*, 612–631.

Warner, M., & Bel, G. (2008). Competition or monopoly? Comparing privatization of local public services in the US and Spain. *Public Administration, 86*, 1–13.

Warner, M., & Hefetz, A. (2002a) Applying market solutions to public services: An assessment of efficiency, equity and voice. *Urban Affairs Review, 38*, 70–89.

Warner, M., & Hefetz, A. (2002b) The uneven distribution of market solutions for public goods. *Journal of Urban Affairs, 24*, 445–459.

Warner, M., & Hefetz, A. (2003). Rural – urban differences in privatization: limits to the competitive state. *Environment and Planning C: Government and Policy, 21*, 703–718.

Yoo, C. S. (2002). Vertical integration and media regulation in the new economy. Yale *Journal on Regulation, 19*, 171–300.

Chapter 7
How to Get Increasing Competition in the Dutch Refuse Collection Market?

E. Dijkgraaf and R.H.J.M. Gradus

Abstract For the refuse collection market, it is well-known that concentration increases prices and offsets the advantage of contracting out. The presence of competing public firms might be essential to ensure fair competition. In this chapter we show that increasing competition by public firms decreases prices and can be essential for low prices.

Keywords Collection · dynamics · concentration · prices · Netherlands

7.1 Introduction

In the nineties, contracting out public services has become an important measure to improve efficiency within the public sector (see for example Savas, 1987). There seems much empirical evidence that especially contracting out refuse collection reduces costs. Domberger and Jensen (1997) conclude that contracting out suggests cost savings of twenty percent. Given these costs advantages the use of private collectors seems scarce. For example, in the Netherlands 38% of the contracts for municipal refuse collection is placed privately (see Table 7.1). For the UK, Netherlands, Sweden and Ireland similar pictures can be given (see Dijkgraaf, Gradus, & Melenberg, 2003).

Therefore, political economy papers have empirically studied the privatization factors of especially refuse collection (see for example López-de-Silanes, Shleifer, and Vishny for the United States, Bel and Miralles (2003) for Spain, Dijkgraaf, Gradus & Melenberg, 2003) for the Netherlands, Ohlsson (2003) for Sweden and

---

✉ E. Dijkgraaf
SEOR-ECRi, Erasmus University Rotterdam, P.O. Box 1738, 3000 DR Rotterdam, The Netherlands
e-mail: dijkgraaf@few.eur.nl

R.H.J.M. Gradus
Faculty of Economics and Business Administration, VU University Amsterdam, De Boelelaan 1105, 1081 HV Amsterdam, The Netherlands
e-mail: rgradus@feweb.vu.nl

E. Dijkgraaf, R.H.J.M. Gradus (eds.), *The Waste Market*,

Table 7.1 Waste collection in the Netherlands in 2002 and 2006

	Observations in 2002		Observations in 2006	
Collection by:	Number	% of municipalities	Number	% of municipalities
– private firm	183	37	174	38
– public firm	88	18	115	25
– municipal[1]	95	19	64	14
– municipal services[2]	130	26	105	23
Total	496	100	458	100

[1]Cooperation with neighbor.
[2]Only own municipality.

Christoffersen and Paldam (2003) for Denmark). The overall conclusion of this literature is that the privatization decision of the municipality is pragmatic and not ideological. Therefore, other explanations are investigated as well. Another explanation is that the market for refuse collection is monopolized. A high degree of market concentration may weaken competition and, therefore, makes it difficult for some local governments to obtain benefits from contracting out. Recently, two empirical papers try to investigate this issue. Based on calculation of the Herfindahl-index for Catalonia, Bel and Warner (2006) argues that a tendency toward concentration may diminish the advantage of contracting out. In Dijkgraaf & Gradus (2007) it is shown that there is a correlation between regional concentration and prices. In highly concentrated provinces competition is weak, which results in higher prices. Moreover, in low concentrated provinces where public firms are active competition is strengthened. Therefore, the paper shows that involving public firms competing for tendering can be a proper tool for reducing costs in potentially concentrated markets. However, one of the disadvantages of this study was that data for the private side of the market was only available for 2002. This not only resulted in estimation results which were not significant at a high level, but also in a lack of information on the price effects of changes in concentration. Therefore, we stressed that an important topic for future research should be contracting out dynamics. There are some indications in the literature that the effects of a special mode of production will change over time. Szymanski (1996) and Bel and Warner (2006) stress out that the advantage of privatization of refuse collection disappears due to a tendency toward monopoly, while Hefetz and Warner (2004) show that there is a reverse privatization trend in the USA.

Therefore, in this chapter we investigate these topics by using observations for two years. By comparing the Herfindahl index for 2002 and 2006 we can describe market dynamics over time and try to investigate the relation between (changes in) market concentration and prices. This chapter is organized as follows. In the second paragraph we calculate the 2002 and 2006 Herfindahl index for the Dutch refuse collection market. It is shown that public firms increased their market shares. In the third paragraph we discuss the methodology of estimating a cost function and the available data. In the fourth paragraph estimations are discussed. Interestingly, the positive relation between concentration and prices is now highly significant. Moreover, we show empirically that the presence of enough competing public firms might be essential to offset the disadvantage of high concentration. Finally, in the fifth paragraph some conclusions are drawn.

7.2 Development of the Dutch Refuse Collection Market Over Time

Dutch municipalities have a legal obligation to provide a waste collection infrastructure for municipal waste. They are free to choose whether to provide this task themselves or to contract out waste collection to outside firms. In the 1990s contracting out occurred only to private firms. As an alternative, municipalities cooperated to vest new public firms. At the start these firms only collected waste for the municipalities that owned the firm. During the last years, however, public firms began to compete with private firms for contracts by other municipalities. Nowadays contracting out to outside firms involves both private and public firms.

In 2002 183 municipalities (i.e. 37% of the municipalities and 26% of the inhabitants) have contracted out waste collection to a private firm and 85 to a public firm (see Table 7.1). A third group of municipalities (95) collects the waste by a municipal service in cooperation with neighboring municipalities. The other municipalities (133) collect the waste themselves (i.e. 26% of the municipalities and 38% of the inhabitants).

Interestingly, the market share of especially public firms increases substantially from 88 municipalities (i.e. 18%) in 2002 to 115 municipalities (i.e. 25%) in 2006 (see also Dijkgraaf and Gradus (2008)). There seems a pattern that public collectors are increasingly a preferred choice for Dutch municipalities. The number of municipalities using a private firm is more or less the same. In 2002, 37% of all municipalities uses private firms and in 2006 38%. The other two institutional forms decrease over time. The share of municipalities collecting the waste themselves decreases from 130 municipalities (i.e. 26%) in 2002 to 105 municipalities (i.e. 23%) in 2006 and municipal cooperation decreases with 31 municipalities (i.e. 5%) from 95 municipalities (i.e. 19%) in 2002 to 64 municipalities (i.e. 14%) in 2006.

In Dijkgraaf and Gradus (2007) we show that the Dutch waste collection market was highly concentrated with respect to competition between private firms in 2002 (see also Table 7.2). For the national market the 2002-Herfindahl index is 0.27. If the relevant market is the province, concentration is even higher. The concentration is not evenly spread over the country. Some provinces do not have private collection firms at all (Flevoland and Friesland), while others have a high incidence of private collection. However, public firms behave more and more as competitors for private firms. On a national scale the 2002-Herfindahl index is now only 11% suggesting a competitive market. Still concentration might be available at a provincial level as for a number of provinces the Herfindahl index is still very high.

In this contribution we want to focus on the development of the Herfindahl index over time. An interesting question is whether private firms have strengthened their market position and whether the role of public firms has changed. Therefore, in

Table 7.2 Overview Herfindahl index 2002 and 2006

	Observations in 2002		Observations in 2006	
	Private competitors	Private and Public competitors	Private competitors	Private and Public competitors
Drenthe	1.00	1.00	1.00	0.56
Friesland	n.a.	1.00	n.a.	1.00
Flevoland	n.a.	0.85	n.a.	0.80
Groningen	0.72	0.71	0.65	0.45
Limburg	0.53	0.50	0.63	0.63
Zuid-Holland	0.43	0.35	0.57	0.26
Zeeland	0.58	0.34	0.55	0.61
Utrecht	0.57	0.32	0.43	0.30
Noord-Holland	0.46	0.32	0.45	0.22
Overijssel	0.23	0.31	0.25	0.34
Noord-Brabant	0.28	0.20	0.22	0.16
Gelderland	0.28	0.16	0.43	0.18
Netherlands	0.27	0.11	0.23	0.08

Table 7.2 the Herfindahl index in 2006 is included at a national and provincial level. In most provinces, the difference between 2006 and 2002 in the Herfindahl index for private firms is small. In other provinces, where a change is visible, institutional developments are important. For example, in Zuid-Holland a large public company, i.e. AVR, has become private. The decrease in Utrecht is due to the mergers of municipalities. So, the 2006-situation on the private side of the market is more or less comparable to 2002. There are two dominant firms in both years, where SITA serves 87 and Van Gansewinkel 39 municipalities.

Compared with the relatively stable private market, the number of public firms has increasing substantially between 2002 and 2006. As a result the Herfindahl index is decreasing in most cases. In some provinces municipal cooperation has changed into a public firm. In Noord-Holland, Holland Collect is a new public firm, which was based on municipal cooperation in the area of West-Friesland. In other provinces public firms were municipal services before. In Drenthe, the public NV Area Reiniging consists of a merger of the municipal services of Coevorden, Emmen and Hoogeveen at January 1 2006. In the province of Groningen, the public firm Omrin has entered the market, while they were only active in Friesland in 2002. In Zeeland, however, an increase is visible due to fact that an existing public firm now collects waste in five municipalities.

7.3 Methodology and Data

We test whether concentration influences refuse collection costs by an OLS estimation of a standard log linear total cost function. This function includes as production variable the number of collection vehicle stops (measured by the number of households) and a number of exogenous factors like the travel time to the pick-up points, the time needed to collect the waste, the waste composition and waste treatment costs (see Dijkgraaf & Gradus, 2007).[1,2] Furthermore, three dummies are included that measure whether the waste collection firm is public, private or an intermunicipal cooperation. Municipalities that collect the waste themselves are the benchmark for these variables. Finally, variables are included that measure regional concentration and competition by public firms. We test four alternatives:

- First, we include Herfindahl indices in the cost function and multiply the Herfindahl indices by the private ownership dummy. As public companies compete with private companies both are included in this variable. We multiply the Herfindahl indices by the private ownership dummy to test the effects of concentrated markets on the behavior of private firms.
- Second, we also include the Herfindahl indices multiplied by the public ownership dummy to test the effects of concentrated markets on the behavior of public firms.
- Third, we capture the dynamics between 2002 and 2006 by taking as an dependent variable the increase (or decrease) of prices between 2002 and 2006 and include the Herfindahl indices multiplied by the private and public ownership dummy as independent variable.
- Fourth, we include the dynamics in Herfindahl indices as well. Hereby, we take again as an dependent variable the increase of prices between 2002 and 2006, but now include the change in Herfindahl indices between 2006 and 2002 both multiplied by the private and public ownership dummy as independent variable.

Data for the type of collection (by the municipality itself, by public firm, by private firm or by an intermunicipal cooperation), waste composition and total costs come from the Dutch Waste Management Council. Total costs are calculated by multiplying the average municipal tariff per household with the number of households per municipality. If actual tariffs do not cover total costs, we use the coverage factors to calculate cost covering tariffs.[3] Other data for exogenous variables come from the Dutch Bureau of Statistics. The same source is used for the number of inhabitants per municipality, the basis for the calculation of the Herfindahl indices. All data are for nearly all Dutch municipalities in 2002 (496) and 2006 (458). In total we have 866 observations as for 43 municipalities in 2002 and 45 in 2006 data are missing. Table 7.3 gives the descriptive statistics for the variables described above.

---

[1] Note that factor prices are not included as no reason is present why they should differ between municipalities.

[2] In the Netherlands waste treatment costs depend on the waste incineration plant. There are 10 plants in the Netherlands, so we include 9 dummies (with the plant of HVC (the plant serving the largest number of municipalities) as a benchmark). Note that in 2002 and 2006 in the Netherlands none of the municipalities use another form of treatment, like landfilling. Furthermore, we have no information that the situation in 2006 is different from the situation in 2002.

[3] As only companies have to pay VAT, we use tariffs excluding VAT for these firms (see Wassenaar & Gradus, 2004).

Table 7.3 Descriptive statistics

	Mean	Max.	Min.	Std. Dev.
Municipal collection costs (million euro)	3.52	114.24	0.11	7.54
Pickup-points (households)	15,115	410,201	490	30,248
Inhabitants per point	2.49	3.65	1.76	0.21
Density (hectares per household)	0.0119	0.4063	0.0004	0.0251
Unsorted waste (kg per household)	221	529	71	59
Glass (kg per household)	23	116	5	8
Paper (kg per household)	73	158	16	16
Vegetable, fruit and garden waste (kg/hh)	109	301	2	44
Collection with neighboring municipalities	0.17	1	0	0.38
Collection by public firm	0.21	1	0	0.41
Collection by private firm	0.37	1	0	0.49
Herfindahl (private and public)	0.37	100	0.16	0.23

7.4 Results

According to the first estimation, private collection is 20% cheaper than collection by municipalities (see Table 7.4).[4] This result is consistent with the literature. Collection by a public firm is 18% cheaper than collection by municipalities. Although the coefficient for private firm collection is somewhat higher than for public firms, a Wald test does not reject the hypothesis that they have the same size. Apparently, the most important factor influencing collection costs is not ownership but contracting out. Moreover, the difference between collection by an intermunicipal cooperation and collection by the own municipality is insignificant.

Table 7.4 Estimation results: effect on total cost waste collection

Independent variable	Model 1	Model 2	Model 3
Effect for municipalities with collection by:			
– intermunicipal cooperation	0.02	0.02	0.02
	(0.02)	(0.02)	(0.02)
– public firm	−0.08***	−0.08***	−0.09***
	(0.02)	(0.02)	(0.03)
– private firm: all	−0.08***	−0.14***	−0.14***
	(0.02)	(0.03)	(0.03)
– private firm: effect of Herfindahl		0.16***	0.17***
		(0.06)	(0.06)
– public firm: effect of Herfindahl			0.03
			(0.05)
R^2-adjusted	0.97	0.97	0.97

Notes: Standard errors beneath coefficients. Coefficients with * /**/*** significant at 90/95/99%.

[4] As the estimations are in logs the effect can be calculated using $e^x - 1$. Note that this effect has to be multiplied by 2.5 as collection costs are on average 40% of total costs. We only present results for production mode and concentration variables. Results for other variables are available upon request.

The second estimation shows that the coefficient of the Herfindahl index is significant at 99%. Interestingly, the positive relation between concentration and prices is now highly significant compared with Dijkgraaf & Gradus (2007)[5], which strengthen our case that the costs of private provision are increasingly dependent on regional concentration. At the average value of the Herfindahl index the net effect of private provision on collection costs is −16%. With a Herfindahl index of 1 (monopoly) total costs even increase with 9%. At the other hand, cost advantages of private collection are much higher if enough competition is present. A Herfindahl index of zero (maximal competition), results in an estimated cost decrease of 32%. This stresses that the competitiveness of the market is extremely important when waste collection is contracted out.

In the third estimation the added coefficient for the Herfindahl indices multiplied by the public ownership dummy is not significant and, therefore, implies no effects of concentrated markets on the behavior of public firms. This is also an interesting result. According to this estimation, it seems that public companies can play an important role, if they compete with private companies. However, there price behavior seems not be influenced by market concentration.

In the fourth estimation (Table 7.5) we explore the market dynamics and investigate whether the increase of prices is related to the institutional dummy or the Herfindahl. For the institutional dummies we find a negative sign meaning that prices go up if self supply is chosen. However, the coefficient for cooperation and private firms are not significant at all. For public firms it is significant at 90%

Table 7.5 Estimation results: effect on total cost waste collection change (2006/2002)

Independent variable	Model 4	Model 5
Effect for municipalities with collection by:		
– intermunicipal cooperation	−0.06	−0.07
	(0.04)	(0.04)
– public firm	−0.09*	−0.34***
	(0.05)	(0.09)
– private firm: all	−0.06	−0.03
	(0.05)	(0.08)
– private firm: effect of Herfindahl	0.08	
	(0.11)	
– public firm: effect of Herfindahl	0.17*	
	(0.09)	
– private firm: effect of change in Herfindahl (2006/2002)		0.17
		(8.44)
– public firm: effect of change in Herfindahl (2006/2002)		0.35***
		(0.10)
R^2-adjusted	0.02	0.05

Notes: Standard errors beneath coefficients. Coefficients with */**/*** significant at 90/95/99%.

[5] For the Herfindahl-index significance at 90%-level is found and for C3-ratio we found significance at 95%-level (see Dijkgraaf & Gradus, 2007).

indicating some evidence. Interestingly, the decreasing effect of public firms on prices is smaller, if market concentration is higher. However, this effect is again only significant at 90%-level. At the average value of the Herfindahl index the effect of public firms leads to 6% lower prices. With a Herfindahl index of 1 (monopoly) the change in total costs is positive (22%). At the other hand, changes in cost of public collection are much higher if enough competition is present. A Herfindahl index of zero (maximal competition), results in 21% higher cost changes.

In the fifth estimation we explore the market dynamics further and investigate whether the increase of prices is related to the change in the Herfindahl index. Interestingly, the results are much stronger. For the public firm dummy we find a significant negative coefficient meaning that prices go down if a public firm is chosen. Similar, the effect of public firms on prices is smaller, if market concentration is higher. Interestingly, this effect is now significant at 99%-level and thus indicates the importance of the presence of enough firms also for public firms. Although the level of concentration does not influence the price of public firms, the change in concentration measured by the Herfindahl index does. At the average value of the change in the Herfindahl index (−9%) the effect is negative, implying a decrease in prices of 5%. If the Herfindahl index increases with 9%, the net effect is 10%. Thus, not only the level, but also the change in Herfindahl index determines price paths.

7.5 Conclusions

In this article we show that the Dutch market for private refuse collection is highly concentrated as the Herfindahl indices for 2002 and 2006 are high on a provincial level. Also if public firms are included the Herfindahl indices stay high. Moreover, it is shown that in highly concentrated provinces competition is weak, which results in barriers for local governments to effectively obtain benefits from contracting out. However, according to our estimates this is only the case for private firms. The price behavior of public firms seems not influenced by market concentration and in low concentrated provinces, where public firms are active, competition is strengthened. The importance of public firms is also put forward, if the increase of prices between 2002 and 2006 is related to the institutional dummy. For public firms we find a significant negative sign meaning that prices go down if a public firm is chosen in a market where competition increases. Thus, the level of concentration does not influence the cost of public firms, but the change in concentration does.

The involvement of public firms seems an effective way to organize day-to-day operations under private commercial law rules, whereas the government retains control over strategic decisions as will be done in a public firm. An important policy implication of this chapter is that local governments should be cautious with privatization of public firms. Although it raises some short run revenues, it can cause welfare losses in the long run.

There are several topics for future research. Although we have data for two years, it would be worthwhile to investigate the issue for a longer panel data set. In the literature it is stressed that the advantage of privatization refuse collection disappears over time due to a tendency toward monopoly (e.g. Bel and Warner (2006)). Therefore, it is important to investigate whether a transformation of a local government division into public-owned private-law cooperation can offset the tendency toward a monopoly dynamically. Finally, an important topic for future research is the relevant market. There are some indications that the relevant market for refuse collection is the province and this assumption is used in the empirical part of this chapter. Till recently the market was regulated and organized on a provincial level. However, current legislation is more on a national scale and in some cases even on an international scale. This stimulates cooperation between regions in different provinces. It would be worthwhile in future research to analyze whether other relevant markets are feasible.

Acknowledgments This chapter has been written when the second author was visiting Universitat de Barcelona and we like to thank especially Germà Bel and Xavier Fageda for stimulating comments.

References List

Bel, G., & Costas, A. (2006). Do public sector reforms get rusty? Local privatization in Spain. *Journal of Policy Reform, 9,* 1–24.

Bel, G., & Miralles, A. (2003). Factors influencing privatization of urban solid waste collection in Spain. *Urban Studies, 40,* 1323–1334.

Bel, G., & Warner, M. (2006). *Local privatisation and costs: a review of the empirical evidence,* working paper Universitat de Barcelona.

Christoffersen, H., & Paldam, M. (2003). Markets and municipalities: A study of the behaviour of Danish municipalities. *Public Choice, 114,* 79–102.

Domberger, S., & Jensen, P. (1997) Contracting out by the public sector: theory, evidence and prospects. *Oxford Review of Economic Policy, 4,* 67–78.

Dijkgraaf, E., & Gradus, R. H. J. M. (2008). Institutional developments in the Dutch waste market. *Environment and Planning C: Government and Policy, 26,* 110–126.

Dijkgraaf, E., & Gradus, R. H. J. M. (2007). Collusion in the Dutch waste collection market. *Local Government Studies, 33,* 573–588.

Dijkgraaf, E., Gradus, R. H. J. M., & Melenberg, B. (2003). Contracting out refuse collection. *Empirical Economics, 28,* 553–570 (reproduced as chapter 3).

Hefetz, A., & Warner, M. (2004). Privatization and its reverse: explaining the dynamics of the government contracting process. *Journal of Public Administration Research and Theory, 14,* 171–190.

López-de-Silanes, F., Shleifer, A., & Vishny, R. W. (1997). Privatization in the United States. *Rand Journal of Economics, 28,* 447–471.

Ohlsson, H. (2003). Ownership and production costs. Choosing between public production and contracting out in the case of Swedish refuse collection. *Fiscal Studies, 24,* 451–476 (reproduced in chapter 4).

Savas, E. S. (1987). The key to better government. New York: Chatham.

Szymanski, S. (1996). The impact of compulsory competitive tendering on refuse collection services. *Fiscal Studies, 17,* 1–19.

Wassenaar, M. C., & Gradus, R. H. J. M. (2004). Contracting out: The importance of a solution for the VAT distortion, *CESifo Economic Studies, 50,* 377–396.

Chapter 8
Dutch Cost Savings in Unit-Based Pricing of Household Waste

E. Dijkgraaf and R.H.J.M. Gradus

Abstract We estimate the effects of four unit-based pricing systems on waste collected in Dutch municipalities. Unit-based pricing is shown to be effective in reducing unsorted and compostable waste and in stimulating recyclable waste. If the estimations are corrected for differences in environmental activism between municipalities the effects are still large but significantly lower. The bag-based and weight-based systems perform equally and far better compared with the frequency-based and volume-based systems. This is interesting, as administrative costs are significantly lower for the bag-based system. Finally, unit-based pricing has no effect on the amounts of waste collected in surrounding municipalities.

Keywords Municipal waste management · unit-based pricing systems · environmental activism

8.1 Introduction

More and more Dutch communities have implemented unit-based user fees to finance waste collection. These user fees require households to pay for each kilogram, bag or can presented at the curb for collection. By 2000, more than 20% of all Dutch municipalities had implemented such a system. In this chapter, we estimate household reactions to the implementation of unit-based pricing for the collection of residential waste. Our estimates show significant and sizable price effects, which depend on the type of unit-based pricing.

✉ E. Dijkgraaf
SEOR-ECRi, Erasmus University Rotterdam, P.O. Box 1738, 3000 DR Rotterdam, The Netherlands
e-mail: dijkgraaf@few.eur.nl

R.H.J.M. Gradus
Faculty of Economics and Business Administration, VU University Amsterdam, De Boelelaan 1105, 1081 HV Amsterdam, The Netherlands
e-mail: rgradus@feweb.vu.nl

E. Dijkgraaf, R.H.J.M. Gradus (eds.), *The Waste Market*,

Table 8.1 Overview of the econometric literature on effects of unit-based pricing

Study	Country	System	Elasticities		
			Price[a]	Own-price[b]	Cross-price[c]
Household surveys					
Hong et al., 1993	USA	Volume	3.63	not sig.	> 0
Van Houtven & Morris, 1999	USA	Volume	0.22	–0.10	not sig.
Jenkins, Martinez, Palmer, & Podolsky, 2003	USA	Volume[d]	2.13		not sig.
Reschovsky & Stone, 1994	USA	Bag (recyclable)	0.85		not sig.
Reschovsky & Stone, 1994	USA	Bag (compost)	0.85		> 0
Fullerton & Kinnaman, 1996	USA	Bag	0.89	–0.08	0.07
Van Houtven & Morris, 1999	USA	Bag	0.86	–0.26	not sig.
Hong, 1999	Korea	Bag	1.49	–0.15	0.46
Linderhof et al., 2001	NL	Weight (compost)[e]	3.86	–1.39	
Linderhof et al., 2001	NL	Weight (unsorted)[e]	4.14	–0.34	
Aggregate municipality data					
Wertz, 1976	USA	Volume	5.85	–0.15	
Jenkins 1993	USA	Volume	1.46	–0.12	
Strathman et al., 1995	USA	Volume	5.69	–0.45	
Van Houtven & Morris, 1999[f]	USA	Volume	0.22	< 0	
Kinnaman & Fullerton, 1997	USA	Bag	0.16	–0.19	0.23
Podolsky & Spiegel, 1998	USA	Bag	3.62	–0.39	
Van Houtven & Morris, 1999[f]	USA	Bag	0.86	–0.15	
Kinnaman & Fullerton, 2000	USA	Bag	0.09	< 0	not sig.
Callan & Thomas, 1997	USA	Mixed[g]	n.a.		0.07

[a]Average tariff in real US dollars (2000) per 30 gallons (114 liters) of unsorted waste.
[b]Elasticity of the amount of collected unsorted waste with respect to the price of unsorted waste collected at the curbside.
[c]Elasticity of the amount of collected recyclable (and/or compostable) waste with respect to the price of unsorted waste collected at the curbside.
[d]Of the 1,049 households, 116 face a positive unit price, of which 104 subscribe to collection of a pre-specified number of cans and 12 pay per bag/tag/sticker.
[e]In Oostzaan, the city Linderhof et al. (2001) study, both compostable and unsorted waste are priced on a weight basis.
[f]Data are aggregated per sanitation route.
[g]In Massachusetts, different unit-based pricing systems exist (bag, tag, volume). This study does not discriminate between the different programs.

Two streams of literature that estimate household reactions to the implementation of unit-based pricing systems can be distinguished. The first uses cross-sectional analyses of municipalities and the second applies household survey data. Most of the studies show considerable impacts from a pricing system. Table 8.1 summarizes the existing econometric literature with respect to the effects of unit-based pricing. In general, nearly all studies find a negative and significant own-price effect from unit-based pricing. The results are more mixed for the cross-price effect on collected recyclable waste.

Most studies evaluate bag- or volume-based systems. Only Linderhof Kooreman, Allers and Wiersma (2001) study the effects of the most refined, weight-based system. Table 8.1 indicates that own-price elasticities overlap for the different unit-based pricing systems. For example, Strathman, Rufolo, and Mildner (1995) found an elasticity of –0.45 for the volume-based system, which is higher than the elasticities of the bag-based systems, while Hong, Adams, and Love (1993) found a non-significant elasticity. Direct comparison of systems is limited to Van Houtven & Morris (1999). This chapter compares the effects of bag- and volume-based systems and finds a significantly higher elasticity for the bag-based system for curbside-collected unsorted waste. The effect on the quantity of waste recycled is found to be insignificant in both cases.

We extend the literature in three directions. Firstly, we explicitly distinguish between the different systems of unit-based pricing (weight-based, bag-based, frequency-based and volume-based pricing). This contributes to the literature because no study presents a direct comparison of the possible unit-based pricing systems. Our results clearly indicate that the bag- and weight-based systems perform far better than the other systems. Secondly, we investigate whether environmental activism is responsible for part of the estimated price effect. Our research shows that municipalities that introduce a unit-based pricing system already produce less waste on average before its introduction. When no correction is made for this effect, price effects estimated on the basis of cross-section data might overestimate the true effects. Thirdly, we test whether surrounding municipalities without unit-based pricing systems in fact collect part of the waste produced in municipalities with unit-based pricing systems. No such effect seems to be present in Dutch municipalities.

8.2 Effects of Unit-Based Pricing

8.2.1 Method and Data

In previous studies using cross-sections of municipalities, waste per capita is a function of price, the municipality's mean level of income, the share of homeowners, the age distribution, the average number of people in a household and other demographic variables (see for example Fullerton & Kinnaman, 1996). We use the quantity of waste collected (in kilograms per inhabitant) also as the dependent variable. However, we are able to discriminate between different waste streams. In the Netherlands, municipalities are obliged to collect three types of waste separately: compostable waste such as vegetable, food and garden waste; recyclable waste such as glass, paper and textiles; and unsorted waste. Furthermore, municipalities are obliged to collect compostable and unsorted waste at the curbside. For recyclable waste, municipalities can choose whether they collect at the curbside or provide drop-off centers.[1] For municipalities without curbside collection of recyclable waste, the number and location of drop-off centers must be such that the collection infrastructure is easily accessible for all citizens. For example, municipalities place collection units at shopping centers and at entrance roads of neighborhoods.

Data on the dependent variables, the quantities collected of total, unsorted, recyclable and compostable waste in kilograms per inhabitant, come from studies by the Dutch Waste Management Council (AOO). Total waste collected is calculated as the sum of unsorted, recyclable and compostable waste. The AOO-studies present data on the quantities of paper, glass, textiles, compostable and unsorted waste collected for 1998, 1999 and 2000. The AOO uses an annual inquiry from the CBS (the Dutch Central Bureau for Statistics), which is sent to the waste collection units of all Dutch municipalities. These units have reliable figures for the quantity of waste collected as the bill they have to pay is based on the quantity of waste supplied to waste treatment firms. These firms weigh the waste each time a collection vehicle brings waste to the treatment plant. The CBS checks the quality of the data by comparison with other years and by comparison with additional information from waste treatment companies.[2] Additionally, as the data for sorted waste are partly collected by schools and charitable organizations, information from regional and national representative organizations for glass, paper and textiles recycling is used to check these data. The response rate of the inquiry is 91%. Thus, our data-set comprises nearly all Dutch municipalities. The actual number of municipalities included differs for each dependent variable due to data availability. The first four rows in Table 8.2 present summary and availability statistics for the dependent variables (see the Appendix for the variable definitions).[3]

Dutch municipalities are free to choose the financing mechanism for waste collection. Most municipalities finance waste collection by a flat rate (see Table 8.3). This results in a marginal price of zero. In order to promote waste prevention and recycling, a number of municipalities have introduced a unit-based pricing system. In general, the Dutch unit-based pricing systems generate marginal prices for unsorted and compostable waste, while the collection of recyclable waste (glass, paper and textiles) is still free. This gives citizens the incentive to sort their waste and to change their buying behavior. Different Dutch municipalities have introduced different types of unit-based pricing systems. These systems can be ordered with respect to the refinement of the pricing system. It could be expected on theoretical grounds that as marginal pricing becomes more and more refined, households respond with greater reductions in priced waste streams and a growing supply of unpriced waste streams.

---

[1] In some municipalities, there is a free curbside collection program for recyclable paper organized by local associations, such as sports clubs and schools. Our data include the waste collected by these associations.

[2] In the inquiry, municipalities are asked which companies treat the waste. Information from these companies is gathered to make comparison possible.

[3] As not for all municipalities data are available for all years, the number of observations is not exactly equal to the number of years multiplied by the number of cross-sections.

Table 8.2 Descriptive statistics

	Mean	Maximum	Minimum	Standard deviation	Number of observations	Number of cross-sections
$Waste_{total}$	431	707	222	62	1.323	507
$Waste_{unsorted}$	218	450	52	54	1.451	530
$Waste_{compost}$	117	239	12	39	1.449	529
$Waste_{recyclable}$	99	217	19	20	1.334	508
UBP_{weight}	0.02	1.00	0.00	0.14	1.451	530
$UBP_{bagunscom}$	0.01	1.00	0.00	0.11	1.451	530
UBP_{baguns}	0.02	1.00	0.00	0.15	1.451	530
UBP_{fre}	0.07	1.00	0.00	0.26	1.451	530
UBP_{vol}	0.05	1.00	0.00	0.22	1.451	530
UBP_{oth}	0.01	1.00	0.00	0.12	1.451	530
Retire	13.31	27.77	6.38	2.90	1.451	530
Fam size	2.56	3.70	1.72	0.20	1.451	530
Foreigner	0.04	0.31	0.00	0.04	1.451	530
City	0.05	1.00	0.00	0.22	1.451	530
Village	0.57	1.00	0.00	0.50	1.451	530
Density	0.50	27.46	0.02	1.35	1.451	530
Ownhouse	10.05	30.59	1.34	3.12	1.451	530
Ownflat	1.68	16.53	0.00	2.20	1.451	530
Income	39.04	44.60	28.50	2.34	1.451	530

Table 8.3 Occurrence of unit-based pricing systems

	1998	1999	2000
Municipalities with unit-based pricing systems			
Weight-based system	9	10	13
Bag-based system for both unsorted and compostable waste	6	6	6
Bag-based system for unsorted waste	13	12	14
Frequency-based system	19	43	54
Volume-based system	24	30	29
Unspecified type of system	6	8	10
Total	77	109	126
Municipalities without unit-based pricing systems	461	429	412
Total	538	538	538

In general, four different systems are present: volume-based, frequency-based, bag-based and weight-based.[4] Table 8.3 gives an overview of the pricing systems used by Dutch municipalities in the period 1998–2000 based on the annual AOO inquiry.

The volume-based program allows households to choose between different volumes of collection can. Most municipalities supply a standard can with a volume of

[4] Some municipalities have a combination of the different unit-based pricing systems or apply the pricing system to only part of their municipality. These are included in Table 8.3 as ‘unspecified type of system’.

140 liters (37 gallons), with the possibility of upgrading to a 240-liter (63 gallon) can or of subscribing to more 140-liter cans. In general, citizens can choose different volumes for unsorted and compostable waste. The marginal price in the volume-based system is rather crude, as the decision on the optimal level of waste supply can only be made at the beginning of the contract period and at certain review times (usually annual). In 2000, 29 municipalities in the Netherlands used a volume-based pricing system.

A more refined marginal price results from a frequency-based system, in which the household pays for the number of times the can is presented at the curbside. The payment is not dependent on the actual amount of waste the can contains. Whether the can is filled or half empty, the bill household receive is just equal to the number of times the can is presented. The occurrence of frequency-based pricing systems shows a notable rise between 1998 and 2000. In 2000, this type of system was the most frequently used pricing system.

In the bag-based system, households have to buy a special bag with specific marks. In most cases, these bags can be bought at supermarkets, petrol stations and the town hall. Other bags without the relevant marks are not collected. The bag-based system is a more refined pricing system than the frequency-based system, as the volume of the bag is significantly less than that of the can. In the Netherlands, the volume of bags is 50 or 60 liters (13 or 16 gallon). An important difference compared with other unit-based pricing systems is that the most frequently used bag-based system leaves compostable waste unpriced. In 2000, 14 municipalities used a bag-based system for unsorted waste in combination with a free collection can for compostable waste. Only a minority of municipalities that have a bag-based system use bags for both unsorted and compostable waste (6 municipalities in 2000). As the incentives of the two systems differ, we include both types separately in the estimations.

Maximum flexibility results from a weight-based system. The collection vehicle weighs the can and combines this information with the identity of the owner, stored in a chip integrated in the collection can. In this case, a greater weight of waste results in a higher collection fee. While the number of municipalities using a weight-based system has increased, in 2000 still only 13 municipalities had introduced such a system.

As data are available for 1998–2000, we estimate a panel model using both the cross-section and the time-related variation.[5] For each waste stream (total waste, unsorted waste, recyclable waste and compostable waste), we estimate:

$$Waste_{w,i,t} = \alpha_s UBP_s + \beta SE + c_i + d_t + \varepsilon_{i,t}, \tag{8.1}$$

[5] We tested the assumption that pooling the different years is valid. An F-test on the sum of squared residuals rejected this assumption at the 99% level (F-statistic is 2.04). However, we only present results for the pooled model because a comparison with results for the separate years showed that the estimated coefficients are very robust. Only for the frequency variable was the coefficient significantly different from the panel estimates at the 95% level for 1998 (–0.11) and 2000 (–0.29). The reason for this is the sharp rise in the number of municipalities using the frequency system.

where $Waste_{w,i,t}$ is the quantity of waste stream w in municipality i in year t, UBP_s are dummies with the value 1 if municipality i has a unit-based pricing system of type s in year t, SE is a vector of socio-economic characteristics, c_i are time-invariant regional fixed effects,[6] d_t are time fixed effects and $\varepsilon_{i,t}$ is the normally distributed error term (where necessary corrected for cross-sectional heteroskedasticity).[7]

To correct for differences between municipalities, we include the following socio-economic characteristics: the area of a municipality per inhabitant (and its square), the average family size, the number of non-western foreigners per inhabitant, the percentage of total inhabitants earning a median income, the number of houses sold per inhabitant, the number of flats sold per inhabitant, a dummy for small municipalities, a dummy for large municipalities and the percentage of inhabitants older than 65.[8] Data for the socio-economic characteristics come from the CBS (the Dutch Central Bureau for Statistics). Descriptive statistics for the variables are given in Table 8.2.

8.2.2 Results

Table 8.4 presents the estimation results. The F-statistics show that the equations are significant, while the relatively high (adjusted) R^2s indicate that the explained variation is not small.

Pricing waste on the basis of weight has a highly negative and significant effect on total waste of 38%.[9] This effect differs for the underlying waste streams. Compostable waste diminishes by more than 60%. It seems that many Dutch households use home composting methods to reduce this type of waste. Also, the effect on unsorted waste, the most environmentally unfriendly waste stream, is large: introducing a weighing system reduces the amount by nearly 50%. From the estimations, it is

---

[6] Ideally, we would include a fixed effect for each municipality. However, as the unit-based pricing system dummies are highly invariant with respect to time, this is not possible. As a second best, we include a dummy for each province. Results for these fixed effects are available upon request.

[7] We tested all specifications for heteroskedasticity using the Breusch-Pagan test. It showed that for estimations with the independent variables in levels, heteroskedasticity could not be rejected. Therefore we estimated with the independent and, where possible, right-hand-side variables in logs (see Appendix). In cases where heteroskedasticity could still not be rejected, we corrected the standard errors with the White procedure (see Table 8.4).

[8] We tested the robustness of the estimated coefficients for the unit-based pricing systems by estimating a wide variety of different equations. Excluding some of the control variables or including extra control variables (such as the percentage of inhabitants in full-time work, the percentage of western foreigners, the number of families with 1, 2 or more children, the amount of property tax paid and the size of the agriculture sector) showed that the estimated coefficients for the unit-based pricing systems are very robust. For example, the coefficients for total waste are between –0.48 and –0.53 for the weight-based system and between –0.23 and –0.26 for the frequency-based systems. Further results are available upon request.

[9] As the dependent variable is in logs, the effects of the pricing dummies are calculated using e^x-1, where x is the estimated coefficient.

Table 8.4 Estimation results: dependent is ln(*Waste*)

	Total	Unsorted	Compostable	Recyclable
UBP_{weight}	–0.48	–0.68	–0.95	0.19
	(0.02)	(0.03)	(0.05)	(0.03)
$UBP_{bagunscom}$	–0.44	–0.68	–0.93	0.26
	(0.02)	(0.04)	(0.06)	(0.03)
UBP_{baguns}	–0.15	–0.74	0.31	0.15
	(0.02)	(0.03)	(0.04)	(0.02)
UBP_{fre}	–0.24	–0.32	–0.46	0.09
	(0.01)	(0.02)	(0.03)	(0.02)
UBP_{vol}	–0.07	–0.13	–0.01#	0.03#
	(0.02)	(0.02)	(0.03)	(0.02)
UBP_{oth}	–0.15	–0.47	–0.02#	–0.01#
	(0.03)	(0.04)	(0.06)	(0.05)
ln(Retire)	0.11	0.04#	0.27	0.09**
	(0.02)	(0.03)	(0.05)	(0.04)
ln(Fam size)	–0.24**	–0.61	0.55	0.31*
	(0.08)	(0.11)	(0.17)	(0.16)
ln(Foreigner)	–0.03	–0.00#	–0.12	–0.02*
	(0.01)	(0.01)	(0.02)	(0.01)
City	–0.05	0.01#	–0.23	–0.15
	(0.01)	(0.02)	(0.04)	(0.03)
Village	0.01#	–0.03**	0.03*	0.05
	(0.01)	(0.01)	(0.02)	(0.02)
ln(Density)	0.03	0.09	0.03*	0.00#
	(0.01)	(0.01)	(0.01)	(0.03)
$(\ln(Density))^2$	0.004*	0.028	–0.016	0.002#
	(0.002)	(0.003)	(0.005)	(0.009)
Own_{house}	0.002**	0.003*	0.015	0.002#
	(0.001)	(0.002)	(0.003)	(0.002)
Own_{flat}	–0.007	0.001#	–0.024	–0.013
	(0.002)	(0.003)	(0.004)	(0.004)
ln(Income)	0.24	0.24	0.07#	0.27#
	(0.06)	(0.09)	(0.14)	(0.17)
R^2 (adjusted)	0.63	0.68	0.63	0.26
F-statistic	77.31	106.85	87.80	17.50
White correction	Yes	No	No	Yes
Fixed effects	Yes	Yes	Yes	Yes
Number of observations	1,323	1,451	1,449	1,334

Notes: Equations are estimated including a constant. Standard errors are given in parentheses. All coefficients are significant at the 99% confidence level, except for coefficients with **(*) which denotes significance at the 95% (90%) level and for coefficients with # which denotes non-significance at the usual levels.

clear that one of the important mechanisms generating this result is that the amount of recyclable waste increases when a unit-based pricing system is introduced: introducing the weight-based system leads to higher efforts in recycling glass, paper and textiles (up 21%). Of course, this is due to the fact that Dutch citizens do not have to pay a marginal price for the collection of this type of waste. Given the cross-price effect, the net decrease in unsorted waste is 29%.

Introducing a bag-based pricing system also reduces the amount of total waste. In municipalities that use the bag-based system both for unsorted and for compostable waste, total waste diminishes by 36%. For municipalities that collect compostable waste by using a free collection can, the reduction is only 14%. While the effects on unsorted waste are comparable for the two systems (–49% and –52%), the effects on the supply of compostable waste differ greatly. In municipalities with unpriced compostable waste collection, compostable waste increases (by 36%), while in the other municipalities (using a bag system for compostable waste as well as for unsorted waste), this waste decreases (by 61%). Interestingly, the effect on recyclable waste is also larger for municipalities that use the bag-based system for compostable waste. This suggests that in municipalities using a bag-based system only for unsorted waste, part of the recyclable waste is 'dumped' in the free compostable waste can. The intuition behind this result is that it takes less effort to use this can than to use the recyclables facility. The compostable waste can is in the direct vicinity of the house, while the collection infrastructure for recyclable waste is farther away, resulting in more time needed to deliver the recyclables. Interestingly, the effects of the bag-based system that prices both unsorted and compostable waste are comparable to those of the weight-based system.

The system based on frequency reduces the total amount of waste by 21%, due to a reduction in both unsorted waste (27%) and compostable waste (37%). As the effects on unsorted waste are less pronounced than in the weight-based and bag-based systems, the stimulating effect on the collection of recyclable waste is smaller as well (up 10%).

The effects of introducing a system based only on the volume of the collection are smaller. Total waste decreases by only 6%, mainly due to the effect on unsorted waste as the effects on compostable and recyclable waste are insignificant. This result is not surprising since the volume-based system is less refined than the other systems.

Turning to the socio-economic characteristics, we find economies of scale for total waste. This corresponds to the results found in the literature. An increase in household size of one standard deviation reduces collected waste per inhabitant by 5%. Diseconomies of scale are found for compostable waste. A possible explanation is that households with three or more people are more likely to have a garden.

In addition, the amount of waste per capita is larger for municipalities with a larger population of elderly people or a smaller population of foreign people. This is especially the case for compostable waste. As the garden area of the household primarily determines the amount of compostable waste, it is clear that living in a city has a highly significant and negative effect on compostable waste and living in a village has a positive effect. Furthermore, as we should expect, the sign on compostable waste is negative for municipalities with many flats. Moreover, a larger area per inhabitant increases the waste stream. The coefficients on income for total and unsorted waste are in accordance with the literature and positive, while income has no influence on compostable and recyclable waste.

8.2.3 The Price Elasticities of the Pricing Systems

So far, we have estimated the effects of unit-based pricing systems using dummies for the different systems, as no information is available on tariffs for 1998–2000. However, we do have data on the tariffs in 2003.[10] Assuming that these tariffs are a proxy for the real tariffs in 1998–2000, we can estimate the price elasticities of the different unit-based pricing systems. This makes comparison with results found in the literature easier.

Table 8.5 presents the estimated elasticities. Consistent with the results presented in Table 8.4, the price elasticities are highest for the weight-based system and the bag-based system that prices both unsorted and compostable waste. This is interesting, as the average tariff for the weight-based system is more than twice that for the bag-based system. The better results for the bag-based system are very clear when the elasticities of the volume and frequency systems are compared. While the average tariff for the volume-based system is more or less equal to that of the bag-based system and the tariff for the frequency system is 1.89 dollars higher, their elasticities are significantly lower.

The higher price elasticity for unsorted waste in the bag-based system than in the volume-based system is in line with the results of Van Houtven & Morris (1999). The much smaller average tariff for the bag based system in their study than in ours might explain the lower own-price elasticity found for the bag-based system and the insignificant effects on recycling compared with our findings.

Table 8.5 Estimated price elasticities

System	Price	Total	Unsorted	Compost	Recyclable
Standard model					
Weight	4.39	–0.47	–0.67	–0.92	0.16
Bag, unsorted + compostable	2.02	–0.43	–0.66	–0.97	0.25
Bag, unsorted	2.15	–0.14	–0.71	0.29	0.14
Frequency	3.91	–0.22	–0.28	–0.40	0.08
Volume	1.94	–0.06	–0.12	–0.01#	0.01#
Model with environmental activism					
Weight	4.39	–0.40	–0.53	–0.81	0.12
Bag, unsorted + compostable	2.02	–0.36	–0.51	–0.85	0.20
Bag, unsorted	2.15	–0.07	–0.58	0.40	0.09
Frequency	3.91	–0.16	–0.16	–0.31	0.04*
Volume	1.94	–0.00#	0.01#	0.09	–0.03#

Note: Equations are estimated including the same socio-economic characteristics as presented in Table 8.4 (results are highly comparable and available on request).

[10] In the estimations we use the tariffs charged each time a can is emptied for the frequency system. For the volume system, we use the marginal weekly increase in the collection fee if a household subscribes to a larger can. To make comparisons between systems possible, the reported tariffs in Tables 8.1 and 8.5 are in real (2000) US dollars (using the GDP deflator) per 30 gallons (114 liters) of unsorted waste. Tariffs per mass unit are transformed to tariffs per volume unit using a regularly reported maximum weight of 0.76 kilograms per gallon (3.79 liters).

Compared with the elasticities found in the literature, our estimated own-price elasticities for the bag-based and weight-based systems are high. For example, the study with the highest elasticity for the bag-based system (Podolsky & Spiegel, 1998) finds an elasticity of only −0.39.

Comparing the average tariffs charged in Dutch municipalities with the average prices charged by communities whose elasticities are estimated in the literature reveals that the average Dutch tariff for the volume-based system is similar to the average tariffs reported in other studies (compare Tables 8.1 and 8.5). The average Dutch tariffs for the frequency-based, bag-based and weight-based systems are inside the range of tariffs evaluated in the literature. Thus, the higher own-price effects we estimated are not the result of higher prices in the Netherlands.

Interestingly, the cross-price elasticities we found for recyclable waste are not outside the range found in the literature. This suggests that the larger effects of bag-based and weight-based pricing in the Netherlands are not the result of more substitution between unsorted and recyclable waste. In the next two sections, we analyze whether the high Dutch elasticities are influenced by citizens' environmental activism and by leakage effects to neighboring municipalities.

8.3 The Importance of Environmental Activism

Section 8.2 shows that unit-based pricing systems have a significant effect on the quantity of collected waste. Part of this effect may, however, result from a higher level of environmental activism. Figure 8.1 illustrates this point. Assume that citizens in municipality B (where unit-based pricing is introduced in the second period) are more concerned about the waste problem than citizens in the flat-fee municipality, A. Our method to estimate the effects of unit-based pricing systems compares the waste quantities of both municipalities, resulting in an estimate that is the sum of the environmental-activism effect and the price-system effect. The true effect of the price system for municipalities with a level of environmental activism comparable to that in municipality B is, however, equal to the difference in the second period

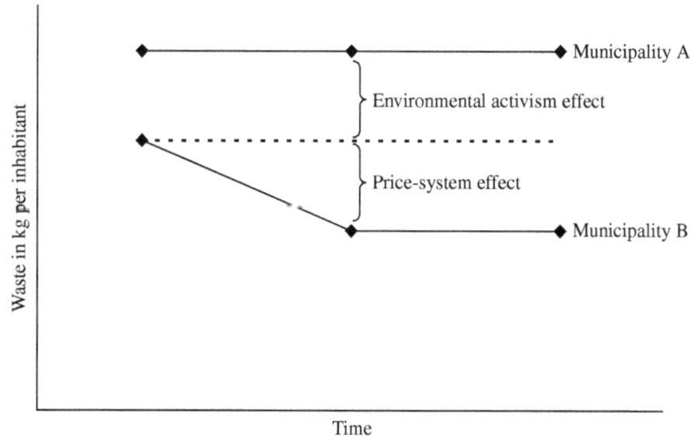

Fig. 8.1 Influence of environmental activism on quantity of waste

minus the difference in the first period. The figures presented in Table 8.4 thus can overestimate the effects of unit-based pricing on the waste quantity of municipalities where such a system is introduced.

A way to deal with the environmental-activism effect is to take into account the political affiliation of the population. For example, Linderhof et al. (2001) suggest that because of the political affiliation of Oostzaan the estimated effects of the weight-based system of Oostzaan may not generalize for other municipalities. They evaluate the introduction of weight-based pricing in this small Dutch city using data before and after introduction of the pricing system. The largest political party in Oostzaan is Green Left (38% of the total vote), which is the most environmentally-friendly-oriented political party in the Netherlands. Green Left received only 7% of the votes nationwide in the parliamentary elections of 1998. This suggests that environmental activism is relatively high in Oostzaan, resulting in less-than-average amounts of waste before the introduction of the weight-based pricing system. Thus, the effect of introducing such a system in municipalities with less environmentally conscious citizens might be larger.

To check the influence of environmental activism, we included the fractions of the vote attained by each political party (based on the local election results of March 1998) in the estimations presented in Table 8.4. The Dutch political parties have different preferences with respect to environmental issues. There is consensus in the Netherlands about the position of most parties on an environmental left-right scale. For example, based on an evaluation of election programs, the Dutch Friends of the Earth gave Green Left an 8 for environmentally friendly policy proposals, while the right liberal party (VVD) was only given a 4.[11] It could be expected that municipalities in which green parties received a high percentage of the votes produce less waste than right-wing municipalities. However, statistical analysis shows that none of the Dutch political parties has a significant influence on the total amount of waste and therefore we conclude that political affiliation is a weak explanatory variable for environmental activism.[12]

Therefore, we check the influence of environmental activism in another way.[13] The communities that most want to recycle and to minimize waste going to disposal might be the ones that choose unit-pricing systems. If so, the pricing system and environmental activism are simultaneously determined with waste quantity. Therefore, the estimated effects of a unit-pricing system might already include the effect of environmental activism. To check this, we test whether municipalities that have introduced a unit-based pricing system in later years (1999 or 2000) already have

[11] See Milieudefensie of April 1998, www.milieudefensie.nl/blad/1998/april98/twverkie.htm.

[12] Some significant effects were found for vegetable, food and garden (VFG), glass, paper and textiles (GPT) and solid waste, but the coefficients are very small. When the liberal party VVD's share of the vote increases by 10% percentage points, VFG waste increases by only 0.6%. While this increase is very small, the effects of other parties are lower still. Results are available upon request. In other research, we found also very weak evidence that political variables influence the institutional organization of refuse collection (Dijkgraaf, Gradus, & Melenberg, 2003).

[13] This paragraph is based on a suggestion of the referee.

lower waste quantities in the years before introduction. We do this by including a dummy variable that has the value 1 for each municipality with a unit-based pricing system in one or more years of our sample and the value 0 otherwise.[14] Including this activism dummy now corrects for the initial lower level of waste due to environmental activism in municipalities that introduce a unit-based pricing system.

As Table 8.6 shows, the activism dummy is significant for all waste streams. The results indicate that municipalities with a high level of environmental activism have 7% less waste. This means that a significant part of the estimated reduction in waste is due to environmental activism and not to the unit-based pricing system. Municipalities with a high level of environmental activism have 13% less unsorted waste, while the amount of compostable waste is 10% lower. As recyclable waste in such 'green' municipalities is 4% higher, households in municipalities with a unit-based pricing system are more active in sorting their waste regardless of the presence of such a system. Correction for environmental activism results in somewhat lower effects for the frequency-, weight- and bag-based systems, while the effect of the volume-based system on total waste is now insignificant. The environmental-activism dummy is also positive and significant for the estimations with tariffs. The

Table 8.6 Estimation results including environmental activism: dependent is ln(*waste*)

	Total	Unsorted	Compostable	Recyclable
Activism	–0.07	–0.13	–0.10	0.04*
	(0.01)	(–0.02)	(–0.03)	(0.02)
UBP_{weight}	–0.42	–0.56	–0.83	0.15
	(0.03)	(0.04)	(0.06)	(0.03)
$UBP_{bagunscom}$	–0.38	–0.55	–0.83	0.22
	(0.03)	(0.05)	(0.07)	(0.03)
UBP_{baguns}	–0.09	–0.62	0.40	0.12
	(0.02)	(0.03)	(0.05)	(0.03)
UBP_{fre}	–0.18	–0.20	–0.37	0.06
	(0.02)	(0.02)	(0.04)	(0.02)
UBP_{vol}	–0.01#	–0.01#	0.08**	–0.01#
	(0.02)	(0.03)	(0.04)	(0.03)
UBP_{oth}	–0.09	–0.35	0.12*	–0.05#
	(0.03)	(0.04)	(0.07)	(0.05)
R^2 (adjusted)	0.64	0.69	0.64	0.26
F-statistic	77.75	108.02	85.99	17.01
White correction	Yes	No	No	Yes
Fixed effects	Yes	Yes	Yes	Yes
Number of observations	1,323	1,451	1,449	1,334

Note: Equations are estimated including the same socio-economic characteristics as presented in Table 8.4 (results are highly comparable and available on request).

[14] We also include a dummy for each different type of unit-based pricing system. As expected, the activism effect is larger for municipalities with weight- and bag-based systems than for those with the other systems. However, as the change over time is not large for the individual systems, we only present results for the systems together.

estimated price elasticities are, on average, 0.13 smaller for unsorted waste, 0.10 smaller for compostable waste and 0.05 lower for recyclable waste (see Table 8.5).

The activism effect may explain part of the differences found in the literature. For example, the results based on household data in Fullerton & Kinnaman (1996) and Linderhof et al. (2001) will not be biased as they result from a comparison of the same households over different time periods. In this case, the environmental-activism effect is automatically excluded from the estimations. In contrast, studies that rely on cross-section analysis may overestimate the effects of unit-based pricing. This might explain why studies based on aggregate municipality data generally find larger elasticities than studies based on household surveys (see Table 8.1).

8.4 The Effect on Surrounding Municipalities

Section 8.2 shows that unit-based pricing has a significant effect on the total amount of collected waste. The estimations suggest that one of the reasons for this result is that more waste is sorted. However, no attention was paid in that section to adverse behavioral effects. One of these effects is that unit-based pricing systems may introduce incentives for citizens to take their waste to municipalities without unit-based pricing systems. It seems logical to suppose that surrounding municipalities experience waste tourism as social contacts (family, friends) can be used to avoid the pricing system. For example, Linderhof et al. (2001) report a study by the city of Oostzaan, which estimates that about 4–5% of waste is taken to surrounding municipalities (which is approximately 13–17% of the reduction in waste prompted by the introduction of a weight-based pricing system).

To test whether municipalities without unit-based pricing systems collect part of the waste produced in surrounding municipalities with unit-based pricing systems, we estimate the models presented in Table 8.4 including impact factors. These factors measure how many inhabitants in surrounding municipalities have an incentive to take their waste to another municipality. Inhabitants of a municipality with a unit-based pricing system with one or more municipalities in their neighborhood without such a system do have an incentive for this behavior. Impact factors are calculated using the following equation:

$$IF_{s,i} = \sum_j \left(\left(1 - \delta D_{i,j}\right) \frac{Inh_j}{Inh_i} S_i \right), \tag{8.2}$$

where $IF_{s,i}$ is the impact factor of municipality i having a unit-based pricing system s, i is a vector of all municipalities, j is a vector of the municipalities with a unit-based pricing system s in the neighborhood of municipality i, δ is a factor between 0 and 1, $D_{i,j}$ is the distance between municipality i and municipality j, Inh_i is the number of inhabitants of municipality i, Inh_j is the number of inhabitants of municipality j and S_i is a dummy with value 0 if municipality i itself has a unit-based pricing system and value 1 if it does not.

The impact factor for municipality i is a function of the distance to and the size of municipalities j (municipalities with unit-based pricing systems). The impact factor is larger when:

1. The distance from a municipality with a unit-based pricing system to a municipality without such a system is smaller. A linear relationship between impact and distance is assumed, while only municipalities with a distance less than 50 kilometers are included, i.e. $\delta = 0.02$ (the impact of municipalities which are more than 50 kilometers away is set to zero). Thus, we assume that taking waste to relatives and acquaintances is less likely if the distance is larger.
2. There are more surrounding municipalities with unit-based pricing systems. If more municipalities with unit-based pricing systems surround a municipality without a unit-based pricing system, the effect will be larger. An extreme example in the Netherlands is Helmond, which does not have a unit-based pricing system and which borders 7 municipalities that have unit-based pricing systems and has 40 municipalities with unit-based pricing systems within a distance of 50 kilometers. On the other hand, 13 municipalities do not have any municipalities with unit-based pricing systems within this distance (consequently, their impact factors are 0). On average, a municipality without a unit-based pricing system has 6 municipalities with unit-based pricing systems in its vicinity.
3. A surrounding municipality with a unit-based pricing system is larger. A surrounding municipality with a unit-based pricing system having the same number as a neighboring municipality without a unit-based pricing system will have less effect on the quantity of waste collected in this latter municipality than will a municipality with 10 times as many inhabitants.

The impact factor is 0 when municipality i itself has a unit-based pricing system. The impact factors are calculated for the different unit-based pricing systems s. For example, $IF_{weight,i}$ is a measure of the impact on collected waste in a municipality without a unit-based pricing system of surrounding municipalities with a weight-based system. Table 8.7 presents the means and standard deviations of the impact factors.

As is shown in Table 8.7, the estimations give little indication of a significant effect from waste tourism. Only 4 out of 20 coefficients are positive and significant, while the size of these coefficients is very small. Furthermore, 3 of the 4 coefficients for the weight-based system are insignificant at 90%, while this system is expected to have the largest effect on surrounding municipalities (evaluated at the mean, the significant effect of the weight-based system is an increase of only 0.6% in the quantity of collected unsorted waste).

To test for misspecification, we also estimated with a non-linear impact factor (decreasing with distance) omitting the scale effect. In this case, only two coefficients are significant. Other estimations also produce few significant coefficients.[15]

---

[15] We estimated models including impact factors calculated with higher ($\delta = 0.013$ and maximum distance of 75 kilometers) or lower ($\delta = 0.04$ and maximum distance of 25 kilometers) influence from neighboring municipalities, impact factors that are only 0 if the same unit-based pricing

Table 8.7 Estimation results: models with impact factors

	Descriptive statistics		Effect of impact variables on ln(Waste)			
	Mean	St. dev.	Total	Unsorted	Compostable	Recyclable
IF_{weight}	0.19	0.44	0.012#	0.028**	0.034#	0.016#
			(0.007)	(0.012)	(0.021)	(0.018)
IF_{bag}	0.86	2.93	–0.001#	0.003#	–0.004#	–0.010*
			(0.001)	(0.002)	(0.003)	(0.005)
IF_{fre}	0.59	2.00	–0.002#	–0.012	0.000#	0.009**
			(0.001)	(0.003)	(0.005)	(0.004)
IF_{vol}	1.02	2.35	0.000#	–0.000	–0.002#	–0.003#
			(0.001)	(0.002)	(0.004)	(0.005)
IF_{oth}	0.29	0.84	0.010	0.007#	0.041	–0.009#
			(0.003)	(0.006)	(0.010)	(0.007)

Note: Equations are estimated including the same socio-economic characteristics as presented in Table 8.4 (results are highly comparable and available on request).

Therefore, we conclude that the taking of waste to municipalities without unit-based pricing systems is relatively unimportant in the Netherlands.

8.5 Administrative Costs and Illegal Dumping

Section 8.2 shows that the effectiveness of bag-based pricing is comparable to that of weight-based pricing. This is an interesting result because the administrative costs for bag-based pricing are much lower. VROM (1997) evaluates weight-, bag- and frequency-based pricing systems in 12 Dutch municipalities.[16] According to this study, average administrative costs are higher for the weight-based pricing system (6.86 euro per inhabitant) than for the other systems (3.18 euro for the bag-based system, 4.28 euro for the frequency-based system).

Given the large reductions in unsorted waste, municipalities can save a lot of money by introducing (especially) a bag-based pricing system. For example, the saving in disposal costs is 5 euro per inhabitant larger than the rise in administrative costs for the bag-based system.[17]

The introduction of unit-based pricing systems may, however, have adverse effects. Citizens may take their waste to neighboring municipalities or may dump

system applies and the environmental-activism dummy. As there was no clear pattern in the results, except that the estimations give insignificant coefficients for nearly all impact variables, we only present the results of estimations with the scale-related linear impact factors with $\delta = 0.02$. Other results are available on request.

[16] The administrative costs for 1997 are given in 2000 prices.

[17] This calculation is based on the cost of incineration (the cheapest available and allowed option in the Netherlands). According to Dijkgraaf, Aalbers, and Varkevisser (2001), total cost per tonne for an efficient incineration plant built in accordance with European law is 77 euro per tonne. Furthermore, VROM (1997) shows that only 2 municipalities (with frequency-based systems) report savings in disposal costs smaller than the rise in administrative costs.

their waste illegally. Analysis of the behavior of Dutch citizens in Section 8.4 shows that there is no evidence that surrounding municipalities without unit-based pricing systems in fact collect part of the waste produced in municipalities with unit-based pricing systems. The evidence on illegal dumping is more mixed. Some studies give support for the hypothesis that illegal dumping is an important issue. Fullerton & Kinnaman (1996) estimate that illegal dumping constitutes 28% of the total reduction in waste collected at the curb. Hong (1999) shows that dumping was substantial after the adoption of the unit-based pricing system in Korea. On the other hand, Reschovsky & Stone (1994) find no relation between illegal dumping and unit-based pricing, while Van Houtven & Morris (1999) report that 'officials ... found little to no evidence of more littering or increased use of accessible dumpsters.'

For the Netherlands, Linderhof et al. (2001) state that illegal dumping is virtually non-existent in Oostzaan. According to them, the monitoring system in Oostzaan, with fines for illegal dumping, appears to be very effective in terms of deterrence. Moreover, another explanation for the absence of illegal dumping is that a small municipality such as Oostzaan has a large degree of social control. In general, the high population density of the Netherlands would suggest a low level of illegal dumping compared with other countries. This is confirmed by the lack of clear anecdotal evidence despite the large number of municipalities with unit-based pricing. However, as the main disadvantage of unit-based pricing systems is the potential effect on illegal dumping, it seems worthwhile investigating an effective monitoring and fining system and the conditions under which such a system would work.

8.6 Conclusions

This chapter provides an empirical analysis of the effects of unit-based pricing of household waste for the Netherlands. We find that the weight- and bag-based pricing systems perform far better than the frequency- and volume-based pricing systems. The bag-based system seems to be the best option, as its effects are comparable to those of the weight-based system and yet its administrative costs are far lower.

Compared with the elasticities found in the literature, the estimated Dutch own-price elasticities for the bag-based and weight-based systems are high. The higher elasticities are not the result of higher marginal tariffs in the Netherlands or of higher cross-price elasticities. A possible explanation might be that more waste is taken to other municipalities (without unit-based pricing systems). However, statistical analysis does not provide evidence that neighboring municipalities do collect part of the waste of municipalities that have unit-based pricing systems. Another possibility is that more waste is illegally dumped. Unfortunately, we have no data with which to estimate the effects on illegal dumping. Monitoring and fining may be important to deter this behavior Given the high population density of the Netherlands and the lack of anecdotal evidence, it seems implausible that a large part of the reduction in unsorted waste is due to illegal dumping.

Therefore, it seems likely that the introduction of unit-based pricing results in a significant change in citizens' behavior. Analysis of the waste quantities before and after introduction of a unit-based pricing system shows that environmental activism does play a role. Waste quantities are lower in municipalities that introduce unit-based pricing in later years. Thus the estimated effects of unit-based pricing may overestimate the effects of unit-based pricing when it is introduced in 'green' municipalities. On average, the estimated price elasticities are 0.13 smaller for unsorted waste, 0.10 smaller for compostable waste and 0.05 lower for recyclable waste when we correct for the environmental-activism effect. However, for municipalities with a low level of environmental activism, the estimated effects based on the dummy-variable approach may be applicable, as introduction of a unit-based pricing system internalizes the lack of environmental activism.

Furthermore, this chapter illustrates that refining unit-based pricing results in greater reductions in collected waste. A simple explanation of why the estimated elasticities for the bag-based system are higher in the Netherlands than elsewhere might be the significantly smaller volume of the bags used (50–60 liters or 13–16 gallons) compared with those in the USA (113–121 liters or 30–32 gallons). That this might be an important issue is indicated by the estimated elasticities of the frequency-based system. While the volume of the Dutch cans in the frequency-based system is comparable to that of the bags in the USA, the estimated Dutch elasticities for the frequency system are also comparable to the elasticities found for the bag program in the USA. Furthermore, the relatively small volume of the Dutch bags might explain why weight-based systems have comparable elasticities.

The smaller bag volume may explain why elasticities for the bag-based system are higher in the Netherlands, but not how Dutch citizens manage to achieve such large decreases in waste as estimated in this chapter. Detailed case studies might be necessary in order to generate enough information to get a grasp of the changes in citizens' behavior when they are confronted with marginal pricing.

Acknowledgments The authors thank an anonymous referee for stimulating comments, the Dutch Waste Management Council (AOO) and Maarten Allers (University of Groningen) for their help in providing some of the data, and Maarten Allers, Loek Bergman (Ministry of Environmental Affairs), Leon Bettendorf (OCFEB) and Peter Kooreman (University of Groningen) for their comments on an earlier version. This chapter was published as Dijkgraaf E, Gradus RHJM (2004) Cost savings of unit-based pricing of household waste: The case of the Netherlands. Resource and Energy Economics 26:353–371.

Appendix: Definition of Variables

$\text{Waste}_{\text{total}}$	Annual total waste collected, in kilograms per inhabitant (sum of unsorted, compostable and recyclable waste) (logged)
$\text{Waste}_{\text{unsorted}}$	Annual unsorted waste collected, in kilograms per inhabitant (logged)
$\text{Waste}_{\text{compostable}}$	Annual compostable waste collected, in kilograms per inhabitant (logged)
$\text{Waste}_{\text{recyclable}}$	Annual recyclable waste (glass, paper and textiles) collected, in kilograms per inhabitant (logged)
$\text{UBP}_{\text{weight}}$	Dummy = 1 if municipality has a weight-based pricing system
$\text{UBP}_{\text{bagunscom}}$	Dummy = 1 if municipality has a bag-based pricing system for both unsorted and compostable waste
$\text{UBP}_{\text{baguns}}$	Dummy = 1 if municipality has a bag-based pricing system for unsorted waste
UBP_{fre}	Dummy = 1 if municipality has a frequency-based pricing system
UBP_{vol}	Dummy = 1 if municipality has a volume-based pricing system
UBP_{oth}	Dummy = 1 if municipality has an unspecified type of pricing system
Retire	Percentage of inhabitants older than 65 (logged)
Fam size	Number of inhabitants per household (logged)
Foreigner	Number of non-western foreigners per inhabitant (logged)
City	Dummy = 1 if municipality has more than 100,000 inhabitants
Village	Dummy = 1 if municipality has less than 20,000 inhabitants
Density	Area of municipality, in hectares per inhabitant (logged)
$\text{Own}_{\text{house}}$	Number of houses sold per 1000 inhabitants
Own_{flat}	Number of flats sold per 1000 inhabitants
Income	Percentage of inhabitants with income over 12,400 and under 21,400 euro (logged)
$\text{IF}_{\text{weight}}$	Impact factor measuring surrounding municipalities with weight-based pricing
IF_{bag}	Impact factor measuring surrounding municipalities with bag-based pricing
IF_{fre}	Impact factor measuring surrounding municipalities with frequency-based pricing
IF_{vol}	Impact factor measuring surrounding municipalities with volume-based pricing
IF_{oth}	Impact factor measuring surrounding municipalities with unspecified type of pricing
Activism	Environmental activism dummy with value 1 for each municipality with a unit-based pricing system in one or more years of our sample and value 0 otherwise.

References List

Callan, S. J., & Thomas, J. M. (1997). The impact of state and local policies on the recycling effort. *Eastern Economic Journal, 4,* 411–423.

Dijkgraaf, E., Aalbers, R. F. T., & Varkevisser, M. (2001). Afvalprijzen zonder grens ("Waste prices without borders"), Studies in Economic Policy 4, Rotterdam: Erasmus University.

Dijkgraaf, E., Gradus, R. H. J. M., & Melenberg, B. (2003). The institutional choice of refuse collection: determining variables in the Netherlands. *Empirical Economics, 28,* 553–570, (reproduced as chapter 3).

Fullerton, D., & Kinnaman, T. C. (1996). Household responses to pricing garbage by the bag. *American Economic Review, 86,* 971–984.

Hong, J. (1999). The effect of unit pricing system upon household solid waste management: the Korean experience. *Journal of Environmental Management, 57,* 1–10.

Hong, S., Adams, R. M., & Love, H. A. (1993). An economic analysis of household recycling of solid waste: the case of Portland, Oregon. *Journal of Environmental Economics and Management, 25,* 136–146.

Jenkins, R. R. (1993). *The economics of solid waste reduction: the impact of user fees.* Aldershot: Edward Elgar.

Jenkins, R. R., Martinez, S. A., Palmer, K., & Podolsky, M. J. (2003). The determinants of household recycling: a material-specific analysis of recycling program features and unit pricing. *Journal of Environmental Economics and Management, 45,* 294–318.

Kinnaman, T. C., & Fullerton, D. (1997). *Garbage and recycling in communities with curbside recycling and unit-based pricing.* NBER Working Paper 6021, Cambridge: NBER,.

Kinnaman, T. C., & Fullerton, D. (2000). Garbage and recycling with endogenous local policy. *Journal of Urban Economics, 48,* 419–442.

Linderhof, V., Kooreman, P., Allers, M., & Wiersma, D. (2001). Weight-based pricing in the collection of household waste: the Oostzaan case. *Resource and Energy Economics, 23,* 359–371.

Podolsky, M. J., & Spiegel, M. (1998). Municipal waste disposal: unit-pricing and recycling opportunities. *Public Works Management and Policy, 3,* 27–39.

Reschovsky, J. D., & Stone, S. E. (1994). Market incentives to encourage household waste recycling: paying for what you throw away. *Journal of Policy Analysis and Management, 13,* 120–139.

Strathman, J. G., Rufolo, A. M., & Mildner, G. C.S. (1995). The demand for solid waste disposal. *Land Economics, 71,* 57–64.

Van Houtven, G. L., & Morris, G. E. (1999). Household behavior under alternative pay-as-you-throw systems for solid waste disposal. *Land Economics, 75,* 515–537.

VROM (1997). *Ervaringen met tariefdifferentiatie en huishoudelijk afval* ("*Experience with differentiated tariffs and domestic waste*"). Den Haag: Ministry of Environmental Affairs.

Wertz, K. L. (1976). Economic factors influencing households' production of refuse. *Journal of Environmental Economics and Management, 2,* 263–272.

Chapter 9
Assessing Instruments for Mixed Household Solid Waste Collection Services in Flanders

X. Gellynck and P. Verhelst

Abstract Instruments to reduce waste can be divided in three groups: first, pecuniary incentives; second, service level; finally, measurements stimulating prevention and waste reduction. Also specific characteristics of the community determine the amount of waste generated. We evaluate whether findings in literature on effectiveness of policy measures are valid for Belgium, specifically for the Flemish region. The policy mix instituted by the Flemish authorities in the 'implementation plan household waste 2003–2007' and implemented by local authorities, is assessed. Multiple regression analysis identifies those measurements having the greatest impact on household solid waste. We found an income elasticity of 0.326. Also the provided service level has a significant impact. Pecuniary incentives are effective instruments in reducing waste, with a price elasticity of –0.139. Furthermore, a higher percentage of direct costs, directly attributable to waste services, borne by households, reduces waste. A consequent implementation of the 'polluter pays' principle proves to be effective.

Keywords Municipal waste management · household solid waste · unit-based pricing

9.1 Introduction

The Sixth Environment Action Programme of the European Commission (EC, 2002) recognizes that a healthy environment is essential to long term prosperity and quality of life. However, future economic development and increasing prosperity will put pressure on the planet's capacity to sustain demand for resources or to absorb pollution. Waste volumes are predicted to continue rising unless remedial action is

✉ X. Gellynck
Department of Agricultural Economics, Ghent University, Coupure links 653, B-9000 Ghent
e-mail: xavier.gellynck@UGent.be

P. Verhelst
Department of Agricultural Economics, Ghent University, Coupure links 653, B-9000 Ghent
e-mail: Pieter.Verhelst@UGent.be

taken. Therefore society must work to de-couple environmental impact and degradation from economic growth. Waste prevention is and will be a key element. The amount of waste for final disposal has to be significantly reduced. Further measures are required to encourage recycling and recovery of wastes.

Council Directive 91/156/EC (EC, 1991) urges the European Member States to take appropriate measures to encourage the prevention and the reduction of waste production and its harmfulness. The recovery of waste by means of recycling, re-use or reclamation or any other process with a view to extracting secondary raw materials or the use of waste as an energy source are promoted. Member States have to take the necessary measures to ensure that waste is recovered or disposed of without endangering human health and without using processes or methods that could harm the environment or the resources of future generations. In order to realize these objectives a waste management plan has to be drawn up, including appropriate measures to encourage rationalization of the collection, sorting and treatment of waste. In the Flemish region of Belgium the government has adopted the 'implementation plan household waste 2003–2007' to comply with the Council Directive (OVAM, 2002).

9.2 Waste Management Instruments

There is a vast literature that can be used to evaluate the effectiveness of instruments used by public authorities aiming at reducing the amount of mixed municipal solid waste, encouraging the amount being recycled and on other factors influencing the amount of waste produced. In the 'implementation plan household waste 2003–2007' mixed household solid waste consists of mixed household waste collected through kerbside collection, bulky household waste and municipal waste such as street-cleaning residues, waste from markets and fly-tipping (OVAM, 2002). This definition is used as case reference point. Assimilated (light commercial) waste is similar in composition to household waste and includes most commercial wastes (Wilson, McDougall, & Willmore, 2001), but is considered in a specific implementation plan (OVAM, 2000). Mixed household solid waste is referred to as waste throughout the chapter.

Table 9.1 summarizes the existing literature with respect to the effects of pecuniary instruments in reducing the amount of waste. The context, including the availability and cost of alternative disposal options, is important to community response to introduction of or changes in the use of pecuniary instruments. The mentioned price and income elasticities have to be interpreted with care.

9.2.1 Pecuniary Incentives

Three groups of studies can be distinguished. A first group of studies focuses on the effectiveness of pecuniary instruments in reducing the amount of waste, going from a fixed annual fee over unit-pricing by the bag to weight-based systems.

Table 9.1 Price and income elasticity estimates of demand for waste collection services

Authors	Data	Model	Dependent variable	Price elasticities	Income elasticities
McFarland, 1972	Cross-section data (1967/68), 13 US cities in California	OLS	Annual per capita quantity of household waste	–0.455	0.178
Cargo, 1976	National survey of community solid waste practices (1968)	OLS			Negative
Wertz, 1976	Cross-sectional data (1970), 16 suburbs Detroit San Francisco (1970)	OLS COM	Annual pounds refuse collected per capita	–0.15	0.272–0.279
Richardson and Havlicek, 1978	Survey and census data (1972/70) Indianapolis, Indiana	OLS	Quantity of kth component in pounds per household per week		0.242
Efaw and Lanen, 1979	Monthly Data (4 years mid 1970s), three US cities	OLS/2SLS	Weight of refuse	Insign.	0.2–0.4
Hong, Adams, & Love, 1993	Self-reported waste quantities, 4306 surveys (1990), Metropolitan Service District, Portland OR	OLS/2SLS	Quantity of household waste	Insign.	Insign.
Jenkins, 1993	Monthly data from 14 towns over several years reported by municipal governments	GLS	Quantity of residential waste discarded per capita per day	–0.12	0.41
Morris and Holthausen, 1994	Simulation Household Production Model	–	–	–0.51–0.60	
Reschovsky and Stone, 1994	Self-reported recycling behavior, 1422 surveys (1990), Tompkins County, NY	Probit	Dichotomous, recycling or not		0.22
Kinnaman and Fullerton, 1997	Cross-section: 959 towns across the US (1991)	OLS			0.262
Podolsky and Spiegel, 1998	Cross-section: 159 towns, New Jersey (1992)	OLS	Demand for household municipal waste disposal	–0.39	0.55

COM=Comparison of means.

Traditional waste management systems charge residents a fixed annual fee for waste collection services. However, this system provides residents with no financial incentive to minimize the total amount of waste they produce. This fixed annual fee can differ between municipalities. McFarland (1972) finds an inelastic price elasticity of demand for waste collection services based on differences in fixed fees of –0.455.

Economic literature devoted to designing waste management policies to achieve the efficient quantity of waste and recycling argues that municipalities should charge according to marginal costs to maximize economic efficiency instead of charging a fixed annual fee. The most direct approach doing is to tax or charge each bag of waste presented by the household. In practice, communities adopting some form of unit-pricing usually turn to average cost pricing that sets the unit-price equal to the average total cost per unit. Several studies describe an inelastic price elasticity of demand for waste collection services in reaction to the introduction of a pay-by-the-bag system. Wertz (1976) estimates a price elasticity of demand for waste collection services of about –0.15. Efaw and Lanen (1979) however find a high inelastic price elasticity of demand for waste collection services, if not perhaps zero or even positive in sign. Later studies confirm the results found by Wertz. Jenkins (1993) finds that pricing waste according to its social marginal cost would reduce the quantity of waste produced by households. A 1% increase in the user fee is estimated to lead to a 0.12% decrease in the quantity of waste. Hong et al. (1993) on the contrary find that a user fee does not appreciably affect the quantity of waste produced at the kerb.

Weight-based fees represent more closely the cost of waste disposal than do volume-based fees, such as unit-pricing by the bag. They also provide a clearer and continuous pricing signal to household producers of waste. Volume based fees provide no additional waste reduction incentive below the lowest level of service, i.e. one bag or bin per collection round (Miranda, Bauer, & Aldy, 1996). The implementation of a price-per-bag program leads to a slight decrease in the weight of waste, but the volume of waste, i.e. number of bags or cans, is characterized by a higher decrease (Fullerton and Kinnaman 1996). Efaw and Lanen (1979) called the observation 'stomping', the changes in user fees can be moderated by the household through volume reduction.

9.2.2 Service Level

A second group of studies also confirms the effectiveness of pecuniary incentives in reducing the amount of waste produced. However, these studies attribute part of the effect to flanking municipal recycling programs. Morris and Holthausen (1994) develop a model of waste decision making for a representative household. They find that households will respond to an increase in unit price by waste reduction and reducing the amount of material recycled and conventionally disposed. Moreover, a joint unit pricing and kerbside recycling program results in substantial welfare improvements to the representative household. The price elasticity of demand for waste collection services is inelastic, between –0.51 and –0.60.

Miranda, Everett, Blume, and Roy (1994) find that the implementation of unit-pricing leads to a reduction in weight of waste by between 17% and 74%. These large estimates cannot be attributed directly to pricing waste, since in every program kerbside recycling programs were implemented during the same year. Also, Podolsky and Spiegel (1998) find a large price elasticity of demand for waste collection services (–0.39) but attribute it in part to the mature recycling programs in place.

Kinnaman and Fullerton (1999) develop a model of household behavior with empirical implications. Households are predicted to respond to an increase in the value of the user fee by decreasing the quantity of waste presented at the kerb. They state that however the implementation of a municipal recycling program diverts some material from waste to recycling, it also frees up additional household resources for consumption, which may result in more waste. Earlier work of Kinnaman and Fullerton (1997) already estimates the impact of kerbside recycling on total household waste, but they find the impact is not statistically significant.

Nevertheless, households may recycle more of the materials that are included in local collection programs. Any increase in recycling presumes that this option is available and that residents find it to be more convenient than disposing of waste through various illegal or undesirable means (Fullerton & Kinnaman, 1996). Recycling attitudes are found to be the major determinant of recycling behavior. These attitudes are influenced by having the appropriate opportunities, facilities and knowledge to recycle (Tonglet, Phillips, & Bates, 2004).

Furthermore, frequency of service can influence the amount of waste collected (Wertz, 1976). Kemper and Quigley (1976) found that the number of collection visits per year is not significantly related to the annual quantity of waste discarded. Platt, Docherty, Broughton, and Morris (1991) and Everett and Peirce (1993) have shown that frequency of recycling collection can have a big influence upon participation and material recovered. The public's perception of waste collection is that it is a system dominated by the collection of waste with an additional recycling service. This needs changing so that the public perceive collection of the recyclable fraction as being the main element of the system. Some authorities are attempting to achieve this by reducing the frequency of refuse collection whilst at the same time increasing the range of the recyclable materials that they collect (Woodard, Bench, & Harder, 2005).

Overall, the disposal services provided are important explanatory factors in the generation of waste. The context, including the availability and cost of alternative disposal options, is important to community response to changes in price and the estimation of any welfare effects associated with changing conditions of service and price (Morris and Holthausen, 1994).

9.2.3 Municipality Specific Characteristics

A third group of studies focuses on specific characteristic of a municipality that can influence the amount of waste produced, as population density and income.

These variables are not in the control of public authorities. Cargo (1976) finds that waste generation is positively correlated with population and density. Dijkgraaf and Gradus (2004) on the contrary find that area per inhabitant increases the waste stream.

As can be seen in Table 9.1, McFarland (1972) determined a small, positive income elasticity of demand for waste collection services of 0.178. Podolsky and Spiegel (1998) find the strongest relationship between waste quantities and income with an income elasticity of demand for waste collection services of 0.55. Other studies also find a positive but weaker relationship between income and waste. Jenkins (1993) estimates an income elasticity of demand for waste collection services equal to 0.41, Efaw and Lanen (1979) between 0.2 and 0.4, Wertz (1976) at 0.279 and 0.272 using two sets of data, Kinnaman and Fullerton (1997) of 0.262, Richardson and Havlicek (1978) at 0.242 and 0.22 by Reschovsky and Stone (1994). While the estimates for income elasticity of demand for waste collection services vary by a factor of almost four, all show waste collection services to be a normal good (Morris and Holthausen, 1994). Hong et al. (1993) estimated a positive but statistically insignificant relationship between waste and the wage rate, which can be seen as a proxy for income. Only Cargo (1976) found a negative correlation between waste generation and income.

9.2.4 Waste Reduction and Prevention

What would induce a household to generate or throw away less waste (source reduction) hinges on at least two elements: the incentive built into the unit-pricing structure for waste collection and disposal, and the availability of convenient (and legal) alternatives such as recycling and yard waste collection or composting programs (Folz & Giles, 2002). One of the major components of household waste is organic material such as kitchen and garden waste, typically comprising 43% by weight of an average household's waste in Flanders (OVAM, 2002) and may include vegetables, fruit, cooked and processed foods, weeds, grass, leafs and other garden waste. Reschovsky & Stone (1994) state that therefore an incentive to participate in composting yard and food waste is likely to generate substantial savings for a locality. One approach that can be adopted by local authorities is to minimize the kitchen and garden waste components of household waste entering the collection stream, through the provision of subsidized waste digesters or compost bins to residents. Home composting has the potential to make a significant contribution to household waste minimization (Bench, Woodard, Harder, & Stantzos, 2005).

9.2.5 Objective

The objective of the chapter is to verify whether the findings in literature are valid for the case of Belgium and more specifically for the Flemish region. In contrast to most other empirical studies the complete municipal policy mix is assessed. Those factors that have the most influence on the amount of waste per capita are identified in order to evaluate the policy choices made. Results are compared with earlier findings in literature.

The structure of the chapter is as follows. Section 9.3 discusses the conceptual framework based on findings in literature. Section 9.4 presents the research methodology and data collection. Section 9.5 sets out the empirical findings and determines the most important factors influencing the amount of mixed municipal solid waste per capita, i.e. mixed household waste, bulky household waste and municipal waste such as street-cleaning residues, waste from markets and fly-tipping. Research findings are compared with results in empirical economics literature. The final section presents conclusions and topics for future research.

9.3 Conceptual Framework

The dominant form of economic analysis has been the assessment of individual instruments instead of some complementary mix. An optimal waste management policy generally requires the simultaneous consideration of instruments (Fenton & Hanley, 1995).

Public authorities have a variety of instruments at their disposal to influence the waste production of households. To what extend they are used by public authorities varies greatly across the world. Even within Belgium there are great differences among municipalities as waste management is a competence of local authorities. The conceptual framework presented in Fig. 9.1 groups the discussed instruments

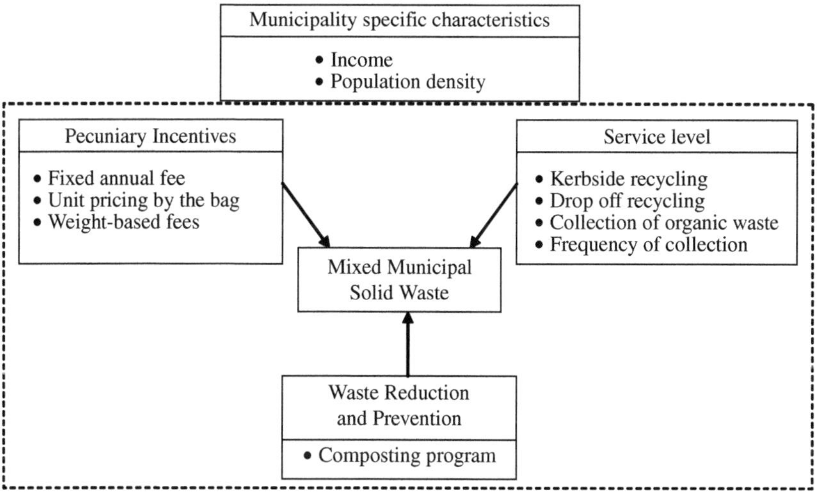

Fig. 9.1 Conceptual framework for evaluating the effectiveness of public policies aimed at reducing the amount of household solid waste

in the previous section in three groups of controllable variables and one group of external variables.

A first group of instruments can be labeled as 'pecuniary incentives' going from a fixed annual fee over unit-pricing by the bag to weight-based fees. The effectiveness of these various forms of pecuniary incentives is demonstrated in economic literature. A second group includes the availability of convenient (and legal) alternatives for waste collection such as kerbside recycling or through a drop-off centre. Frequency of collection and organic waste collection also have a significant influence on the amount of waste produced. These variables describe the service level provided by the public authorities. A third group comprises measurements stimulating people to reduce or prevent waste, having an influence on the amount of waste collected. Promoting yard and food waste composting is a key element because of the important share of this waste fraction in total waste. A fourth group includes specific characteristics of the community not in control of public authorities that also determine the amount of waste. Income and population density are the most discussed parameters influencing waste production.

The conceptual framework gives a basis for evaluating the effectiveness of public policies aimed at reducing the amount of waste produced. In the present research we focus on the Flemish 'implementation plan household waste 2003–2007' that aims at stimulating source reduction and recycling (OVAM, 2002).

Flanders is the northern part of Belgium, situated between the Netherlands and France, bounded by the North Sea. It has a population of 5.9 million, i.e. a population density of 434 inhabitants/km^2. It is subdivided in 308 municipalities (NIS , 2003). The Flemish Parliament and the Flemish Government are fully competent to work out environmental policy. However, their competence for waste management is only directive, through the 'implementation plan household waste 2003-2007'. The management of waste is a task of the municipalities (OVAM 2002).

Delegating disposal authority to the municipalities has resulted in a wide variety of policy approaches. The municipalities decide on fees, receptacles and type and frequency of collection. To encourage the municipalities to adhere to the objectives and instruments put forward in the implementation plan, the Flemish authorities introduced a subsidy plan with significant subsidies for early adopters. The subsidy is diminished through the years.

The main objective of the 'implementation plan household waste 2003–2007' is to reduce the mixed household solid waste to 150 kg/capita/year by 2007, coming from 191 kg/capita in 2000. Mixed household solid waste consists of mixed household waste collected through kerbside collection, bulky household waste and municipal waste such as street-cleaning residues, waste from markets and fly-tipping (OVAM, 2002). A variety of instruments are put forward to realize the objective. The 'polluter pays' principle is selected as the adequate economical instrument to reach this goal. A price of 1.50 euro per bag of 60 l is recommended in order to encourage citizens to reduce the amount of waste. This system is combined with a kerbside recycling program and the installation of drop-off facilities in each municipality, making it easy for citizens to reduce waste through recycling.

The effectiveness in reducing the amount of waste, by the policy mix instituted by the Flemish authorities and implemented in varying degrees by local authorities is analyzed using multiple regression analysis. Results of the Flemish situation are compared with findings in economic literature.

9.4 Methodology and Data Collection

The data sample comprises all 308 municipalities in the Flemish region, fully competent for waste management. Nine coastal municipalities are left out the final analysis. Coastal tourism creates an extra amount of waste in a few months. Specific actions such as a higher frequency of recycling collection during high season or specific receptacles for recyclables on camping sites are set up to tackle this problem (OVAM, 1998). Three other municipalities are also excluded as outliers as their amount of mixed household solid waste/capita is more than three standard deviations higher than the mean value for all municipalities. Another municipality did not deliver sufficient information and consequently is left out. This results in 295 municipalities for final analysis.

Several categories of data are required to make the instruments, put forward in the conceptual framework as described in previous section, operational.

The amount of mixed municipal solid waste/capita for 2003 for each municipality, as dependent variable, is obtained from the regional competent authority, OVAM. Data on instituted pecuniary incentives and characteristics of waste management programs in place for 2003 for all 308 municipalities were obtained on the basis of a structured questionnaire. First, data was gathered from municipalities' web sites and brochures on waste management. Then the data for each municipality was verified and completed via telephone interviews with the competent public officials. Among the categories of information sought are the use of a flat rate fee, a pay-by-the-bag system or a weight-based fee, the type of mixed solid waste and recyclable collection methods utilized both kerbside and drop-off, the existence of composting programs and descriptive data on waste legislation.

Data on municipality characteristics required assessing the factors that influence waste generation, are obtained from the National Institute of Statistics (NIS). Included variables are average income/capita, the area of a municipality per inhabitant and the number of companies. The last variable is included because mixed household solid waste comprises mixed waste from small commercial and industrial activities that cannot be separated from household waste because of its small amount. We hypothesize that a municipality with a high percentage of companies, in particular SME's, is likely to have a larger amount of waste because of this, although assimilated waste is considered in a specific implementation plan and is collected and measured separately (OVAM, 2000). The incorporation of this variable in the regression allows controlling for this effect.

The quantity of collected mixed household solid waste (in kg/capita) is used as the dependent variable in a stepwise multivariate regression analysis in order to determine the relevant factors that influence the amount of waste. Quantities of waste may be affected by a variety of factors, which are controlled for in the multivariate regression analysis. The OLS method is commonly used to measure the effect of pecuniary incentives on the amount of waste, controlling for contextual factors, as is clear from a historical literature review in Table 9.1. In order to comply with standard research practice and to be able to compare findings on income and price elasticity for the Flemish region with earlier findings in other regions we also decided to use the OLS method for our data analysis.

We test for a variety of independent variables, divided into four groups, in accordance with the conceptual framework.

The first group of variables describes the non-controllable characteristics of the municipalities: average income/capita, the area/capita and the number of companies.

A second group of variables describes how the 'polluter pays' principle is implemented, i.e. the pecuniary incentives: flat user fee, yearly cost of kerbside mixed waste collection, presence of a weight-based fee and the percentage of direct costs in total costs of the waste management program for a representative household.

The third group of variables describes the provided service level, i.e. the waste management program in place. Data is collected on the number of waste fractions collected through the kerbside recycling program, the number of fractions collected through drop-off recycling, the frequency of collection visits and the presence of kerbside collection of organic waste.

The last group of variables controls for composting programs through measuring the number of compost masters[1] per 1000 inhabitants and the cost of a compost bin.

The estimated equation is:

$$\begin{aligned} \mathrm{MSW_i} =& \alpha_0 + \alpha_1 \mathrm{AVGINC_i} + \alpha_2 \mathrm{AREA_i} + \alpha_3 \mathrm{COMP_i} + \alpha_4 \mathrm{FEE_i} + \alpha_5 \mathrm{COST_i} \\ &+ \alpha_6 \mathrm{WEIGHT_i} + \alpha_7 \mathrm{PERDIR_i} + \alpha_8 \mathrm{CREC_i} + \alpha_9 \mathrm{DOREC_i} + \alpha_{10} \mathrm{FREQ_i} \\ &+ \alpha_{11} \mathrm{ORGANIC_i} + \alpha_{12} \mathrm{CMASTER_i} + \alpha_{13} \mathrm{PCBINi} + \mathrm{e_i} \end{aligned} \qquad (9.1)$$

where:

MSW = mixed household solid waste per capita in kg, 2003

a. Municipality specific characteristics:
   AVINC = Average income per capita in 1000 euro
   AREA = area per capita in km^2
   COMP = number of companies

b. Pecuniary incentives:
   FEE = flat user fee, 0 if none, in euro
   COST = yearly cost of kerbside waste collection for a representative household in euro

[1] Volunteers that followed a course on composting and are willing to learn other people on how to compost best.

WEIGHT = 1 if weight-based pricing, 0 otherwise
PERDIR = percentage of direct costs in total costs of the waste management program for a representative household

c. Provided service level:
CREC = the number of waste fractions collected through the kerbside recycling program
DOREC = the number of fractions collected through drop-off recycling
FREQ = 1 if once-a-week mixed waste collection, 0 if fortnightly waste collection
ORGANIC = 1 if kerbside collection of organic waste, 0 otherwise

d. Waste prevention and reduction
CMASTER = number of compost masters per 1000 inhabitants
PCBIN = the cost of a compost bin in euro

e. = error term

Summary statistics of the regression variables appear in Table 9.2 .

Table 9.2 Summary statistics of the regression variables

Variable	Mean	Std. Deviation
MSW	146.81	35.60
AVINC	26.30	3.31
AREA	3258.44	2592.79
COMP	473.68	956.43
FEE	65.68	26.38
COST	51.37	19.59
PERDIR	0.68	0.23
CREC	21.68	6.21
DOREC	7.19	1.12
CMASTER	0.43	0.38
PCBIN	15.60	8.75
WEIGHT	94.2%	5.8%
FREQ	56.6%	43.4%
ORGANIC	40.0%	60.0%

9.5 Results and Discussion

The results from the stepwise multiple regression analysis are shown in Table 9.3. Five variables have a statistically significant independent effect on the amount of waste disposed per capita. They explain 37.6% of variance in the production of waste. Over 60% of variance is explained by 'other' variables, indicating the influence of the diversity in the instituted waste management policies by local authorities. There is a high degree of recycling in the Flemish region, up to 70% in 2002. But

Table 9.3 Determinants of mixed municipal solid waste production, 2003

	Unstandard. coefficients	Standardized coefficients			
Independent variables	B	·	t	Sig.	VIF
Constant	131.456		8.285	0.000	
FREQ	26.216	0.370	6.080	0.000	1.621
COST	–0.396	–0.217	–4.332	0.000	1.094
ORGANIC	–20.984	–0.294	–4.669	0.000	1.738
AVINC	1.822	0.169	3.154	0.002	1.253
PERDIR	–0.159	–0.105	–2.108	0.036	1.084
$R^2 = 0.376$					

the recycling programs differ greatly among municipalities. Pecuniary incentives to stimulate recycling, provided recycling service level and accompanying initiatives stimulating waste reduction and prevention for different types of recyclables must be taken into account. The relationship between differences in instituted recycling programs and the amount of waste collected is a topic for further research. For now, only the number of different recycling programs was considered.

The model is tested for multicollinearity. The Variance Inflation Factor (VIF) for all independent variables in Table 9.3 is smaller than 1.8, which is below the tolerance level 2.5 or even 3.

Residuals are dispersed randomly throughout the range of the estimated dependent. Therefore the model gives no violation of homoscedasticity.

As economic literature suggests, waste production is positively correlated with income, both on micro and macro level (see Table 9.1). The model results in Table 9.3 indicate that as the annual average income of people in a municipality increases by 1000 euro, the waste collected increases with 1.822 kg/capita. This gives an income elasticity of demand for waste collection services at the means of the data for mixed municipal solid waste collection of 0.326. This result is in line with most findings in economic literature on income elasticity of demand for waste collection services in industrialized countries with a mature recycling program in place (Podolsky and Spiegel, 1998; Kinnaman and Fullerton 1997; Reschovsky and Stone, 1994; Jenkins, 1993).

Other municipality characteristics such as population density and number of companies have no significant impact on the amount of mixed municipal solid waste collected. The contradicting findings of Cargo (1976) and Dijkgraaf and Gradus (2004) on the impact of population density are not confirmed. The Flemish region with a population density of 434 inhabitants/km^2 is in fact an urbanized area. Only 22 out of 308 municipalities, accounting for 10% of total territory, are rural i.e. with a population density less then 150 inhabitants/km^2 (NIS , 2003; OECD, 1994). Differences between rural areas and urban areas are too small to have a significant impact. The hypothesized influence of number of companies is not significant.

Earlier findings discussed in economic literature demonstrate the effectiveness of ‘pecuniary incentives’, going from a fixed annual fee (McFarland, 1972) over unit-pricing by the bag (Fullerton and Kinnaman, 1996) to weight-based fees (Miranda et al. 1996), in reducing the amount of waste collected. The model results in Table 9.3 show no significant influence from a fixed annual fee, or from the implementation of a weight-based fee. But the implementation of a weight-based fee is made operational as a dummy-variable. Significance of dummy-variables cannot be calculated for in regression analysis. An independent samples T-test in Table 9.4 gives a significant lower average amount of waste collected in municipalities with a weight-based fee in place than in others ($p = 0.000$). The assumption of equal variances however, is violated ($p = 0.066$) which makes the t-test possibly unreliable. So, the data do not allow accepting that weight-based fees have a significant negative impact on waste collected.

However, the level of unit-pricing by the bag, translated in a yearly cost of kerbside waste collection for a representative household has a significant impact on the amount of waste collected. As the yearly cost of kerbside mixed waste collection increases by 1 euro, the waste collected decreases by 0.396 kg/capita (Table 9.3). Controlling for all other variables, a price elasticity of demand at the means of the data for waste collection services of –0.139 is found, which is in line with findings of Wertz (1976) and Jenkins (1993).

Furthermore, the percentage of direct costs borne by households has a significant influence on the amount of waste collected. Direct costs can be directly attributed to the waste services provided, i.e. a fixed annual fee, if any, the costs associated with kerbside collection of waste or recyclables and costs of dropping of recyclables at a drop-off centre. Indirect costs are other general municipal taxes paid by households with no direct link to waste collection. Indirect taxes are used to cover municipal expenses not covered by direct taxes. This result implies that a consequent implementation of the 'polluter pays' principle is an effective instrument in reducing the amount of waste. As the percentage of direct costs increases, the waste collected decreases by 0.159 kg/capita (Table 9.3).

Next to pecuniary incentives, the service level provided has a significant impact on the amount of waste collected (Wertz, 1976; Platt et al. 1991; Everett and Peirce, 1993; Morris and Holthausen, 1994; Woodard et al., 2005). The analysis finds a significant impact from differences in the number of collection visits. A once-a-week collection of waste yields higher amounts of waste than a fortnightly collection round. A higher collection frequency makes it easier to throw waste away than to recycle. Throwing waste away is least time and space consuming compared to recycling. A high frequency of collection prevents other negative side effects, as there are foul odors, problems with vermin and lack of space. A high fee per bag or container does not offset this (OVAM, 2002).

However, significance of dummy-variables cannot be calculated for in regression analysis. But an independent samples T-test in Table 9.4 gives a significant lower average amount of waste collected in municipalities with a every two weeks collection round in place than in others ($p = 0.000$).

Furthermore, the analysis shows that the implementation of a kerbside collection program for organic waste has a significant negative impact on the average amount of waste generated. The organic fraction, both yard waste and waste from fruit and

Table 9.4 T-test testing significance for dummy variables weight, freq and organic

Independent variable		N	Mean	t-value	Sign
WEIGHT	Weight-based fee	17	117.65	–3.547	0.000
	Unit-pricing	278	148.59		
FREQ	Once-a-week	128	167.68	10.261	0.000
	Fortnightly	167	130.81		
ORGANIC	Kerbside collection	177	134.77	–7.810	0.000
	No kerbside collection	118	164.88		

vegetables, comprises about 40% by weight of municipal solid waste in Flanders (OVAM, 2002). However, significance of dummy-variables cannot be calculated for in regression analysis. An independent samples T-test in Table 9.4 gives a significant lower avera ge amount of waste collected in municipalities with a kerbside collection program for organic waste in place than in others ($p = 0.000$).

On the other hand, ease of recycling, measured by the number of waste fractions that are collected in the kerbside recycling program or the number of fractions that can be delivered to the drop-off facility, do not have a significant impact on the amount of waste presented at the kerb (Table 9.3). All municipalities in Flanders have a mature recycling program for paper, glass bottles, plastic and metal containers and drinks cartons (OVAM, 2002). Impact of recycling programs for other fractions is small or insignificant because they ask more effort to recycle.

Finally, next to pecuniary incentives and the service level provided, measurements aimed at stimulating people to prevent or reduce waste can have a significant influence (Reschovsky and Stone, 1994; Bench et al., 2005).

But, efforts to reduce the amount of organic waste presented at the kerbside through stimulating composting do not have a significant impact. Neither the number of compost masters, trained to advise citizens on how to compost, nor the subsidization of a compost bin, stimulates waste reduction in such a way as to reduce waste production significantly (Table 9.3).

9.6 Conclusions and Future Research Topics

Waste volumes are predicted to continue rising unless remedial action is taken (EC, 2002). European Member states should take appropriate measures. The amount of waste for final disposal must be significantly reduced (EC, 1991). Literature on environmental economics evaluates the effectiveness of the instruments used by public authorities to reduce the amount of waste and encouraging the amount being recycled.

The variety of instruments at the disposal of public authorities can be divided in three groups: first, pecuniary incentives, going from a fixed annual fee over unit-pricing by the bag to weight-based fees; second, the provided service level, i.e. availability of kerbside recycling or through a drop-off centre and organic waste collection; finally measurements to stimulate prevention and waste reduction. Besides the instituted policy mix, specific characteristics of the community, as there are income and population density, determine the amount of waste.

The dominant form of economic analysis has been the assessment of individual instruments instead of some complementary mix. Our research however, starts with the identification of the instituted policy mix in each municipality, gathering data for the different aspect of the three groups of policy instruments, supplemented with data on municipality specific characteristics. This allows the identification of factors of the policy mix that have the greatest impact on the amount of mixed municipal solid waste, controlling for all others.

Our findings are in line with earlier findings in literature (see Table 9.1). The higher the annual average income of people in a municipality, the higher the amount of waste. We found an income elasticity of demand for waste collection services of 0.326. This result is in line with most findings in economic literature on income elasticity of demand for waste collection services in industrialized countries with a mature recycling program in place (Podolsky and Spiegel, 1998; Kinnaman and Fullerton, 1997; Reschovsky and Stone, 1994; Jenkins, 1993). However, this variable is out of control of local authorities. The provided service level has a significant impact. A fortnightly collection of waste yields lower amounts of waste than a once-a-week collection round does. The implementation of a kerbside collection program for organic waste can further reduce the amount of waste collected. On the other hand, ease of recycling does not significantly influence the amount of waste collected, due to the mature recycling program in all municipalities for the most common recyclables, such as paper, glass bottles, plastic and metal containers and drinks cartons. Also efforts to reduce the amount of organic waste presented at the kerb do not have a significant impact (Table 9.3).

As demonstrated at large in economic literature, pecuniary incentives, in particular the price of a waste bag, are effective instruments in reducing the amount of waste (see Table 9.1). We found a price elasticity of demand for waste collection services of –0.139 which is in line with findings of Wertz (1976) and Jenkins (1993). Furthermore, the analysis shows that a higher percentage of direct costs, directly attributable to waste services, borne by households, reduces the amount of waste collected (Table 9.3). A consequent implementation of the 'polluter pays' principle, not only for mixed household solid waste collection, proves to be an effective instrument.

Finally, over 60% of variance is explained by 'other' variables, indicating the influence of the diversity in the instituted waste management policies by municipalities, especially for less common recyclables, such as wood, plastics, metals, construction and demolition waste, batteries, tires, oils and fats, small hazardous waste and electronic equipment. The relationship between differences in instituted recycling programs for those recyclables and the amount of waste collected is topic for further research. Pecuniary incentives to stimulate recycling, provided recycling service level and accompanying initiatives stimulating waste reduction and prevention for different types of recyclables must be taken into account.

Acknowledgments This work was carried out at Ghent University, department of agricultural economics. The study was financed by the OVAM, the regional competent authority for waste management. The authors would like to acknowledge the anonymous referees for their many constructive comments and suggestions that helped us to improve the quality of this chapter. This chapter was published as Gellynck X, Verhelst P (2007) Assessing instruments for mixed household solid waste collection services in the Flemish region of Belgium. Resources, Conservation and Recycling 49:372–387.

References List

Bench, M. L., Woodard, R., Harder, M. K., & Stantzos, N. (2005). Waste minimization: home digestion trials of biodegradable waste. *Resources, Conservation and Recycling, 45*, 84–94.

Cargo, D. R. (1976). *Solid wastes: factors influencing generation rates*. Research paper No. 174. Chicago: University of Chicago.

Dijkgraaf, E., & Gradus, R. H. J. M. (2004). Cost savings in unit-based pricing of household waste: The case of the Netherlands. *Resource and Energy Economics, 26*, 353–371 (reproduced in chapter 8).

Efaw, F., & Lanen, W. N. (1979). *Impact of user charges on management of household solid waste*. Report prepared for the U.S. Environmental Protection Agency, Mathtech, Inc., Princeton.

Everett, J. W., & Peirce, J. J. (1993). Curbside recycling in the USA: convenience and mandatory recycling. *Waste Management and Research, 11*, 49–61.

European Commission (2002). Decision no 1600/2002/EC of the European Parliament and of the Council of 22 July 2002 laying down the Sixth Community Environment Action Programme. OJ L 242, 10/9/2002:1–15.

European Commission (1991). Council Directive 91/156/EEC of 18 March 1991. OJ L 078, 26/03/1991:32–37.

Fenton R., & Hanley, N. (1995). Economic instruments and waste minimization: the need for discard-relevant and purchase-relevant instruments. *Environment and Planning A, 27*, 1317–1328.

Folz, D. H., & Giles, J. (2002). Municipal experience with 'pay as you throw' policies: findings from a national survey. *State and Local Government Review, 34*, 105–115.

Fullerton, D., & Kinnaman, T. C. (1996). Household responses to pricing garbage by the bag. *American Economic Review, 86*, 971–984.

Hong, S. H., Adams, R. M., & Love, H. A. (1993). An economic analysis of household recycling of solid wastes: the case of Portland, Oregon. *Journal of environmental Economics and Management, 25*, 136–146.

Jenkins, R. R. (1993). *The economics of solid waste reduction*. Hants, England: Edward Elgar Publishing Ltd.

Kemper, P., & Quigley, J. M. (1976). *The economics of refuse collection*. Cambridge: Ballinger.

Kinnaman, T. C., & Fullerton, D. (1999). *The economics of residential solid waste management*. NBER Working Paper 7326, Cambridge: NBER.

Kinnaman, T. C., & Fullerton, D. (1997). *Garbage and recycling in communities with curbside recycling and unit-based pricing*. NBER Working paper 6021, NBER, Cambridge.

McFarland, J. M. (1972). Comprehensive studies of solid waste management. In: Comprehensive studies of solid waste management, final report, sanitary engineering research laboratory, college of engineering and school of public health, report No. 72–3, (pp. 41–106). Berkeley, CA: University of California.

Miranda, M. L., Bauer, S. D., & Aldy, J. E. (1996). *Unit pricing programs for residential municipal solid waste: an assessment of the Literature*. Report prepared for the U.S. Environmental Protection Agency, Durham: School of Environment, Duke University.

Miranda, M. L., Everett, J. W., Blume, D., & Roy, B. A. (1994). Market-based incentives and residential municipal solid waste. *Journal of Policy Analysis and Management, 13*, 681–698.

Morris, G. E., & Holthausen, D. M. (1994). The economics of household solid waste generation and disposal. *Journal of Environmental Economics and Management, 26*, 215–234.

NIS (2003). *Bevolking en huishoudens. Totale en Belgische bevolking* (*Population and households.* Total and Belgian population). Brussels: NIS.

OECD (1994). *Creating Rural Indicators*, Paris: OECD.

OVAM (2002). *Uitvoeringsplan huishoudelijke afvalstoffen 2003–2007* (*Implementation plan household waste 2003–2007*). Mechelen: OVAM.

OVAM (2000). *Uitvoeringsplan gescheiden inzameling bedrijfsafval van kleine ondernemingen* (*Implementation plan separated industrial waste collection of small enterprises*). Mechelen: OVAM.

OVAM (1998). *Kustactieplan* (*Coastal Action Plan*). Mechelen: OVAM.

Platt, B., Docherty, C., Broughton, A. C., & Morris, D. (1991) Beyond 40 percent: Record setting recycling and composting programs. Washington.

Podolsky, M. J., & Spiegel, M. (1998). Municipal waste disposal: unit-pricing and recycling opportunities. *Public Works Management and Policy, 3*, 27–39.

Reschovsky, J. D., & Stone, S. E. (1994). Market incentives to encourage household waste recycling: Paying for what you throw away. *Journal of Policy Analysis and Management, 13*, 120–139.

Richardson, R. A., & Havlicek, J. (1978). Economic analysis of the composition of household solid wastes. *Journal of Environmental Economics and Management, 5*, 103–111.

Tonglet, M., Phillips, P. S., & Bates, M. P. (2004). Determining the drivers for householder pro-environmental behaviour: Waste minimization compared to recycling. *Resources, Conservation and Recycling, 42*, 27–48.

Wertz K. L. (1976). Economic factors influencing households' production of refuse. *Journal of Environmental Economics and Management, 2*, 263–272.

Wilson, E. J., McDougall, F. R. & Willmore, J. (2001). Euro-trash: Searching Europe for a more sustainable approach to waste management. *Resources, Conservation and Recycling, 31*, 327–346.

Woodard, R., Bench, M., & Harder, M. K. (2005). The development of a UK kerbside scheme using known practice. *Journal of Environmental Management, 75*, 115–127.

Chapter 10
Final Comments and Future Research

E. Dijkgraaf and R.H.J.M. Gradus

Abstract In this chapter we give some final comments and formulate some policy conclusions. Moreover, we give some topics for future research.

Keywords Model specification · data analysis · policy options

10.1 Final Comments

In the previous chapters we analyzed the waste market in different EU-countries. Based on empirical research for the Netherlands and Sweden, we showed that there is little evidence for a positive link between cost savings and privatization. In some cases, such as small sized municipalities, lower costs are reported if waste collection is contracted out. There are, however, reasons to doubt this result in a more general way. The use of private collectors seems relatively scarce, despite the estimated cost advantages at short term. In the Netherlands, for instance, 34% of the municipalities used private firms in 2005. Therefore, political economy models have empirically studied the factors that induce or deter privatization. In this book such an exercise has been done for Dutch and Swedish data. This shows that there is some but not very strong evidence for political patronage. Indeed, a more general finding of the political economy literature is that there seem not many ideological biases influencing politician's decision. The decision of the municipality is often pragmatic and not ideological.

✉ E. Dijkgraaf
SEOR-ECRi, Erasmus University Rotterdam, P.O. Box 1738, 3000 DR Rotterdam, The Netherlands
e-mail: dijkgraaf@few.eur.nl

R.H.J.M. Gradus
Faculty of Economics and Business Administration, VU University Amsterdam, De Boelelaan 1105, 1081 HV Amsterdam, The Netherlands
e-mail: rgradus@feweb.vu.nl

E. Dijkgraaf, R.H.J.M. Gradus (eds.), *The Waste Market*,

Another explanation for the reservations of local authorities toward contracting out is that competitive tendering is no long term viable option due to the monopolistic market structure. We give some evidence that collusion exists in the Dutch and Spanish waste collection market. Private markets in both countries are highly concentrated and, therefore, there are barriers for local government to effectively obtain benefits from contracting out. Interestingly, these countries take different approaches to deal with competition costs. In Spain, where the average size of a municipality is small, intermunicipal cooperation is used to deliver refuse collection. In the Netherlands many public firms bid for contracts outside their jurisdiction(see also Bel, Dijkgraaf, Fageda, & Gradus, 2008). Also Norway uses municipal cooperation to take advantage of economies of scale in the refuse collection market. However, for Norway it is shown that dispersed public ownership as a consequence of municipal cooperation impairs efficiency. Moreover, based on Dutch data it was shown that the price behavior of public firms is not influenced by market concentration and in low concentrated regional markets, where public firms are active, competition is strengthened. From a policy perspective there are at least two important conclusions. First, there is a clear role for the government to promote anti-trust policies in refuse collection markets. Second, local governments should be cautious with privatization of public firms.

Moreover, we analyzed an important innovation in paying for waste costs. In the Benelux and the Scandinavian countries different forms of variable charging based on weight, bag, volume and frequency are becoming more widespread and seem very effective in reducing the amount of waste. With a unit-based pricing (UBP) system the waste collection tariff depends on the amount of waste citizens produce and therefore citizens can reduce the amount of money they have to pay by optimizing the way they sort their waste and my minimizing the waste content of their buying behavior. If they only have to pay for unsorted waste, they have a larger incentive to separate paper, glass and textiles as these components are collected still free of charge. So, also from an environmental point of view this system is profitable. Based on Dutch panel data it is shown that especially the weight- and bag-based system are very effective in reducing waste. In addition, we studied the Flemish policy mix, which combines pecuniary incentives with instruments stimulating prevention and waste recycling. It is shown that the effectiveness of UBP methods can be encouraged if both instruments are combined. Therefore, two other important policy conclusions can be drawn. First, a consequent implementation of serious UBP systems proves to be effective. Second, UBP is more effective if it is combined with local measures stimulating prevention and reduction (see also Dijkgraaf & Gradus, 2007).

10.2 Future Research

In future research there are many avenues to explore. First, it is well-known from the earlier contracting out literature that contracting out refuse collection results in 15–20% lower costs for municipalities. Research in Chapter 2 based on 1996 data for the Netherlands confirms these results. This book shows, however, that this result is not stable over time. It seems that the cost advantage of private provision is much larger in the nineties than at the beginning of this century. In particular, since 2004, costs of municipalities with private provision rise significantly (see also Dijkgraaf and Gradus, 2008). In our view there are two reasons to explain this result. The first explanation is that contracting out refuse collection is a dynamic process typically converging from a competitive market structure to a monopolistic one. For municipalities with a long duration of the mode of private or public production the cost advantage deteriorates over time. Therefore, in future research it is important to study the bidding process in more detail. Interestingly, a study by Gómez-Lobo and Szymanski (2001) shows for the UK that in many biddings the number of bidders is too small for competitive efficiency. The second explanation is based on higher tariffs for public companies due to the introduction of the VAT compensation fund in 2003. Since then they have to pay VAT. As public companies are important competitors for private companies, this VAT introduction might increase the reference price on the market. In addition, studying the institutional effects such as fiscal aspect is an important topic as well to understand market behavior.

Second, we have no indication that our market concentrations results are specific for the Netherlands and Spain. Most mechanisms we are investigated will apply to other countries as well. For other countries there is probably also a tendency for concentration in this market. The literature (e.g. Boyne, 1998; Bel & Warner, 2006) suggests that doubt about the long term efficiency gain of contracting out is not a specific Dutch or Spanish result. However, a specific characteristic of the Netherlands is the presence of public companies and a specific characteristic for Spain is municipal cooperation. As these institutional forms are not present in some other countries, such as the USA and UK, it is worthwhile to analyze whether collusion is a larger problem in other countries (see for example Warner & Bel, 2008). If this is the case, it seems obvious to explore possibilities to increase the scope of institutional possibilities.

Third, the introduction of UBP systems, which seems very effective in reducing waste, may have adverse effects as well. Citizens may dump their waste illegally. Not much evidence, however, is present for this effect in the Netherlands as is shown in Chapter 8. Studying the effects of a weight-based system in Oostzaan, Linderhof, Kooreman, Allers, and Wiersma (2001) state that illegal dumping is virtually non-existent. The monitoring system in Oostzaan, with fines for illegal dumping, appears to be very effective in terms of deterrence. Moreover, a small municipality such as Oostzaan has a large degree of social control. We did not find an effect on neighboring municipalities that have no UBP system. Still, as illegal dumping is the most important factor against UBP, it would be worthwhile to have more sound evidence whether this is a problem, what circumstances influence this problem and which solutions are effective.

Fourth, it is important to reckon with the correlation between the use of contracting out and UBP. If municipalities that contract out use also UBP on a larger scale, previous estimation results of cost functions are biased in favor of contracting out. In addition, it may be hypothesized that the same UBP system is more effectively implemented in municipalities that contract out waste collection and municipalities who choose the most optimal type of waste collection will also be more cost effective in implementation of UBP by reducing administrative costs, better information to households and optimization of the collection infrastructure. In Dijkgraaf and Gradus (2008) we give some evidence for this correlation and correct the previous results in literature. However, in future research it is important to understand more intensively how municipalities can reduce the waste streams and how they can make use of knowledge and experience of external methods.

Also more effort should be paid toward data collection. For the Netherlands, only recently yearly data on refuse collection costs per municipality are available. Based on panel data, contracting out dynamics can be analyzed (Chapter 7) and for the Netherlands there is some evidence that competition deteriorates over time. However, data for more years should be available to fully understand contracting out dynamics. Nevertheless, for other countries such administrative dataset are often not available. In Sweden cost data are available from a survey carried out by the Swedish Competition Authority in 1989 for 115 municipalities. In most other countries, such as Spain and Norway, no cost data are available, although these data are important to understand the cost advantage. Of course, there are costs collecting these data, but these costs are very small compared with the large welfare gains of good policy proposals based on proper data analysis.

In addition, two streams of literature that estimate household reactions to the implementation of UBP systems can be distinguished. The first applies household survey data and the second uses cross-sectional analyses of municipalities. For the first, one gathers data from and with the knowledge of households, which can be hard for a long time-period and can create potential self-selection problems. Nevertheless, if these data become available and the problem of self-selection can be ignored, these data can give rich information on the effect of UBP systems.

More attention is also needed for the implications of model specification. If better data become available, more complex models can be verified as well. In this book we already showed that estimation results can be biased when data for different institutional forms are pooled. In Chapters 2 and 4 we pooled the data and estimated separate costs functions. It was shown that the results confirm our conclusions so far. However, in our empirical analysis we assume for example a Cobb-Douglas production technology and empirical findings in other sectors show that this specification might be too restrictive (see for example De Witte and Dijkgraaf (2007) for an application of the much more flexible Fourier approach for the drinking water sector). Therefore, it would be worthwhile to test alternative cost specification as well. In addition, parametric tests should analyze the robustness of the used model with respect to the specification and its variables. In Chapter 3 we already presented some preliminary results for the Netherlands. Although the same qualitative results are found for parametric and semiparametric models, we found strong statistical evidence that a parametric specification is far too inflexible. Thus, future research is needed for the implications of model specification.

References List

Bel, G., Dijkgraaf, E., Fageda, X., & Gradus, R. H. J. M. (2008). *Similar problems, different solutions: comparing refuse collection in the Netherlands and Spain*. working paper, Universitat de Barcelona.

Bel, G., & Warner, M. (2006). *Local privatisation and costs: a review of the empirical evidence*. Working paper, Universitat de Barcelona.

Boyne, G. A. (1998). Bureaucratic theory meets reality: Public choice and service contracting in US local government. *Public Administration Review, 58,* 474–484.

De Witte, K., & Dijkgraaf, E. (2007). Mean and Bold? Tinbergen Institute Discussion Paper, TI 2007-092/3. Erasmus University Rotterdam.

Dijkgraaf, E., & Gradus, R. H. J. M. (2008). Institutional Developments in the Dutch Waste Market. *Environment and Planning C Government & Policy*, *26*, 110–126.

Dijkgraaf, E., & Gradus R. H. J. M. (2007). Niet bij financiële prikkels alleen. *ESB, 92*, 246–248.

Gómez-Lobo, A., & Szymanski, S. (2001). A law of large numbers: Bidding and compulsory competitive tendering for refuse collection contracts. *Review of Industrial Organization, 18*, 105–113.

Linderhof, V., Kooreman, P., Allers, M., & Wiersma, D. (2001). Weight-based pricing in the collection of household waste: the Oostzaan case. *Resource and Energy Economics*, *23*, 359–371.

Warner, M., & Bel, G. (2008). Competition or monopoly? Comparing privatization of local public services in the US and Spain. *Public Administration, 86*, 1–13.

Index

MIX
Papier aus verantwortungsvollen Quellen
Paper from responsible sources
FSC® C105338

If you have any concerns about our products,
you can contact us on
ProductSafety@springernature.com

In case Publisher is established outside the EU,
the EU authorized representative is:
Springer Nature Customer Service Center GmbH
Europaplatz 3, 69115 Heidelberg, Germany

Printed by Libri Plureos GmbH
in Hamburg, Germany